Contents

List of Tables and Figures

**URANIUM
INSTITUTE**

The Uranium Equation

BALANCE OF SUPPLY AND DEMAND 1980–95

The Supply and Demand Committee
April 1981

The views expressed in this paper do
not necessarily represent those
of any Government with which
individual members of the
Institute may be associated.

Published for The Uranium Institute by
MINING JOURNAL BOOKS LTD.

Published 1981 by Mining Journal Books Ltd.
© *The Uranium Institute*
ISBN 0 900117 27 3
Text set in 11 lens/12 pt Times Medium Roman by photo-composition and printed offset-litho by Chandlers (Printers) Ltd., Bexhill-on-Sea, East Sussex, England.

Corporate Members of the Uranium Institute

(as at July 31, 1981)

Australia	Energy Resources of Australia Ltd. Pancontinental Mining Ltd. Western Mining Corp. Ltd.
Belgium	Synatom SA
Canada	Denison Mines Ltd. Eldorado Nuclear Ltd. Rio Algom Ltd. Saskatchewan Mining Development Corp. Uranerz Canada Ltd.
Europe	Euratom Supply Agency
Federal Republic of Germany	Nukem GmbH Rheinisch-Westfälisches Elektrizitätswerk Aktiengesellschaft Saarberg-Interplan Uran GmbH Urangesellschaft mbH & Co. KG
France	Commissariat à l'Energie Atomique Compagnie Française de Mokta Compagnie Minière Dong Trieu Comurhex Electricité de France Eurodif SA Minatome SA Pechiney Ugine Kuhlmann
Italy	Ente Nazionale per l'Energia Elettrica Ente Nazionale Idrocarburi

Corporate Members of the Uranium Institute — *continued*

Japan	C. Itoh & Co. Ltd.
	The Japan Atomic Power Company
	Kanematsu-Gosho Ltd.
	The Kansai Electric Power Co. Inc.
	Marubeni Corporation
	Mitsubishi Corporation
	Mitsui & Co. Ltd.
	Nichimen Co. Ltd.
	Nissho-Iwai Co. Ltd.
	Power Reactor & Nuclear Fuel Development Corp.
	Sumitomo Corporation
	The Tokyo Electric Power Co. Inc.
South Africa	Atomic Energy Board
	Buffelsfontein Gold Mining Co. Ltd.
	Electricity Supply Commission
	Vaal Reefs Exploration & Mining Co. Ltd.
South West Africa/ Namibia	Rössing Uranium Ltd.
Spain	Empresa Nacional del Uranio SA
Sweden	Luossavaara-Kiirunavaara AB
	Swedish Nuclear Fuel Supply Co.
Switzerland	Gruppe der Kernkraftwerkbetreiber und -projektanten
United Kingdom	British Nuclear Fuels Ltd.
	Central Electricity Generating Board
	The Rio Tinto-Zinc Corp. Ltd.
	Urenco Ltd.
	United Kingdom Atomic Energy Authority
United States	Florida Power & Light Co.

Introduction

In February 1979, the Uranium Institute, through its Supply and Demand Committee, prepared a report on the balance of uranium supply and demand to 1990.* Since then a number of major influences have combined to alter conditions in the uranium market.

The most significant of these is that forecast rates of global economic growth have continued to be progressively lowered, primarily owing to the impact of rising oil costs, which sharply accelerated upwards in the two years to 1981. Many of the previous expectations of Governments and utilities are now invalidated. This economic down-turn has not only restrained the demand for electricity — which is closely coupled with the level of economic activity — but has also had the effect of reducing the sense of urgency.

There has been a noticeable tendency for power plant construction and licensing lead times to increase. Longer lead times discourage, in particular, the construction of nuclear generating capacity because of its relatively high capital cost. From 1980 onwards, high interest rates also aggravated the capital cost problem, including the financing of fuel stocks. Despite this, nuclear power still has a significant overall economic advantage. However, because it now forms a substantial proportion of new planned capacity, nuclear power is particularly affected when cut-backs have to be made.

Governmental control over electricity prices has, in some countries, tended to exacerbate the unavoidable economic effects of high interest rates. It is

---

The Balance of Supply and Demand 1978–1990, published by Mining Journal Books Ltd.

always a temptation for Governments to delay tariff increases which would be well justified by rising costs, in order to limit the impact on the cost of living. There are cases where Public Utility Commissions have prevented utilities from passing on certain costs to electricity consumers, including costs resulting from delays. These distortions of the market economy inevitably reduce the utilities' financial capacity to pursue vigorous investment programmes.

An important factor has been the incident in March 1979 at Three Mile Island. Press and television coverage, highly emotional reporting, and official over-reaction to the incident, led the public at the time to believe that the incident was far more serious than it was in reality. The resulting adverse public reaction towards nuclear power plant construction was not confined to the United States. It was widely felt that "proof" of improvements in nuclear design and safety should be demonstrated before further commitments were made. This public reaction led to temporary closures of reactors already in operation, and to licensing delays to plant under construction, until inspections and adjustments to regulatory procedures could take place.

A further long-term political effect of the Three Mile Island incident has been that it has combined with certain aspects of the policy of the previous U.S. Administration on nuclear non-proliferation to sow in the minds of the U.S. public a suspicion of nuclear fuel reprocessing and waste disposal — both of which are developed techniques that will be needed in any large-scale nuclear power programme. The effects have since been felt far beyond the boundaries of the United States; moreover, they are not confined to the "back end" of the fuel cycle, since delay over the development of waste storage facilities can in some countries form a basis for objectors to question the initial reactor licensing procedures, or even bring a national nuclear programme to a halt.

While it is no exaggeration to say that the Three Mile Island incident has been used as a convenient "flagship" by a number of anti-nuclear political factions, it would be wrong to ignore the positive effects of the incident. By drawing attention to the need for reviewing licensing and operating procedures in the U.S., it could, in the long term, have a beneficial effect in contributing to a greater sense of confidence in the safe operation of nuclear power.

However, some of the effects of these various pressures on the electricity supply industry have been to the direct disadvantage of the public. For instance, some utilities have had to make greater use of relatively expensive oil-fired capacity, or have had to delay the decommissioning of older, less efficient plant, in order to offset the delayed start-up of nuclear reactors. These consequences of the often highly-political campaigns which have been waged against nuclear power deserve to be more widely appreciated.

Finally, the delays in new reactor licensing and construction, and the existence of large utility stockpiles, have had a marked effect in lowering the level of uranium demand, and thus on the uranium mining industry.

This, then, is the background to this second report by the Institute's Supply and Demand Committee. In preparing it the opportunity has been taken for extending the period under review to 1995, and for examining more closely certain aspects which were dealt with only briefly in the 1979 report. The report is based on information available up to the end of April 1981.

In making its forecasts the Institute has been given generous assistance by many of its members, who have brought their collective judgement to bear on the all-important estimates of future installed nuclear capacity and potential production capability. This being, in effect, a report on the future of the industry by the industry itself, there is confidence that the estimated three-fold increase in installed nuclear capacity between 1980 and 1990, shown in this report, is realistic and achievable under current conditions. While, for the reasons mentioned, the expected growth rate is lower than it could have been if there had been fewer constraints, it is still one with which many other industries would be most content.

Demand

Introduction
This section examines both the theoretical demand implied by reactor capacities, and the possible range of real demand within the Western World.* The latter reflect the various policy options available to the operators of reactors and of other components in the nuclear fuel cycle.

Analysis of theoretical uranium demand involves considering the detailed technical and economic operating parameters of the nuclear fuel cycle. The current mix of reactor types is known, and hence the average amount of uranium required under given conditions to generate each unit of electricity; nevertheless, this mix could alter in the future. This report therefore considers the sensitivity of uranium consumption to possible future installed capacity levels and to corresponding reactor mixes. The relatively dependable *committed capacity* is defined, and the scope for results higher than the present *most probable* estimate is considered. Reprocessing of spent fuel and recycling of uranium and plutonium are unlikely to influence demand substantially before 1995, and are not, therefore, taken into account in this report.

Reactor Programmes
The rate of placing new reactor orders has fallen dramatically since 1973, and the decline has been aggravated by cancellations of previous orders, to a point

---

*This is used as a convenient term for those countries, principally the non-centrally planned economies, for which statistics are relatively freely available. This report considers the uranium production of such countries, and the demand arising from reactors they have constructed or are likely to build, for their own use or for export. Thus the report excludes the uranium demand of reactors supplied by the U.S.S.R. to Finland, and includes the demand of Western reactors which may be installed in Yugoslavia and Romania.

4

where the total capacity of existing orders being cancelled slightly exceeded that of new orders being placed from 1978 to 1980. These cancellations reflected, mainly, reduced expectations of the growth in demand for electricity.

The rate of cancellations has slowed in the twelve months since early 1980, and it is reasonable to suggest that orders could soon exceed cancellations. However, the timing of this turnround will depend on the satisfactory resolution of problems relating to the financing and licensing of nuclear power facilities, and to public acceptance of their environmental impact. Some support may be expected from the relative economics of nuclear power: nuclear-generated electricity is considerably cheaper than that obtained from fossil-fuelled facilities in regions having no low-cost coal production. Estimates of the full cost of nuclear-generated electricity in such regions range from 55 to 90% of that for coal-fired generation, based on power stations planned to commence commercial operation in 1990. Oil prices could affect decisions indirectly through their influence on the price of coal: any large coal price rises stimulated by oil shortage would tend to increase the relative attraction of nuclear power.

Figure 1 shows the rates of reactor ordering and cancellation in the U.S. and elsewhere. Orders in the U.S. peaked in 1973 at 52 GWe, dropped to the low level of 5 GWe in 1975, and drifted down to zero by 1979. Cancellations in the U.S. have fluctuated around a level of about 10 GWe annually since 1974, showing continued adjustments to lower growth expectations.

In contrast, the level of ordering outside the U.S. has been fairly constant at around 20 GWe per year since 1971. Cancellations have been at a much lower level, and mostly limited to 1975, 1978 and notably 1979, when the Iranian reactors were cancelled.

Installed capacity projections

In the forecasts which follow, the possible future range of uranium demand is indicated by supplementing the forecast *most probable* installed capacity with indicative upper and lower bounds. These are not intended to be taken as absolute maxima and minima; but they do reflect a relatively wide range of capacities consistent with possible and identifiable national nuclear policies, ranging from a feasible world-wide reactivation of nuclear programmes to a global moratorium. All the capacity projections draw on information and comments supplied by consumer members of the Institute. The information is summarised in short reports on the current status of, and prospects for, nuclear power in individual countries (see Appendix II, page 42). For countries where the Uranium Institute has no consumer member, use has been made of published information.

With lead times for reactor construction and licensing in the range of six to ten years, the installed nuclear capacity up to about 1987 can be predicted with considerably greater confidence than capacity beyond that year. The reactors

5

concerned are already either operating, under construction or on order; and the main uncertainties are confined to construction and licensing lead times. Resolution in some countries of regulatory and financing difficulties would permit a faster build-up of this operating and committed capacity; while further drawing out of current lead times would of course have the opposite effect.

At the lower bound of the forecasts, Table 1 shows how *currently operating* or *committed nuclear capacity* will enter or remain in commercial service. It assumes either no new orders, or that further reactor cancellations will be exactly balanced by new orders. In the case of Japan, reactors approved by the Electrical Power Development Coordination Council are included, these being broadly comparable to reactors on order in other countries. The data for the U.S. reflect the construction and licensing delays currently experienced, so that the *committed capacity* is not shown as fully in place until 1995. Small declines in the committed capacity by 1995 reflect the decommissioning of relatively old reactors.

Once the basis of forecasting is broadened to include reactors not yet on order, ie after 1987, far greater uncertainties are unavoidable. Any estimates which depend on the rate of new reactor ordering also depend on future economic and political developments. The uncertainties are not necessarily all in the downward direction. Indeed, what may be the greatest uncertainty derives from the fact that political initiatives could be taken to shorten lead times for reactor construction and licensing in countries where at present they are overlong. A case in point is the U.S., which currently accounts for 45% of Western World nuclear generating capacity, and which in 1995 could still account for between 35% and 40%.

The forecast *most probable* installed capacity from 1987 to 1995 is given in Table 2. This forecast is based partly on a judgement of each country's past achievements and likely opportunities in relation to the implementation of its nuclear programmes. For the U.S. the estimate is markedly conservative, it being assumed that only the currently operating and committed capacity will be in place. This is equivalent to saying that there will be no new orders or cancellations affecting capacity before 1995, or that these will cancel each other out.

The Institute's projection of a *high growth* nuclear capacity scenario, reflecting achievable national reactor programmes and the potential for benefi-cial political initiatives in a climate of higher economic growth, is detailed in Table 3.

For some countries the *high growth* (Table 3) and *most probable* (Table 2) capacities show little difference. A few countries may be approaching the currently perceived optimum capacity dictated by economic considerations. Others, however, fall considerably short of this economic maximum. In some cases, the differences between Tables 2 and 3 reflect not only different

FIGURE 1. REACTOR ORDERS AND CANCELLATIONS

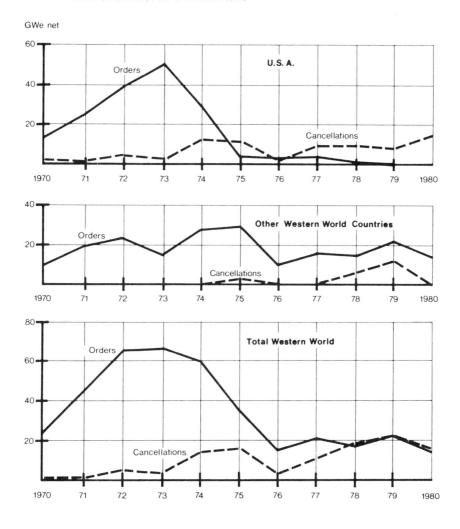

economic growth assumptions, but also judgements regarding the potential for a shift to a more positive climate of public opinion and for corresponding regulatory changes; and of the likelihood, given current capacity levels, that the relevant nuclear programmes can be attained.

While for many less developed countries it is still too early for the published reactor programmes to be assessed in the light of operating experience, the capacity data shown in Tables 2 and 3 are certainly very low in relation to their total energy needs. It is hardly possible to make reliable predictions of the installed capacity of such countries, particularly from 1995 onwards. In preparing Tables 2 and 3 efforts have been made to make some allowance for

7

TABLE 1. OPERATING AND COMMITTED NUCLEAR GENERATING CAPACITY
(GWe, net, year-end, thermal reactors)

	1980	1981	1982	1983	1984	1985	1986	1987	1988	1989	1990	1995
Argentina	0·3	0·3	0·3	0·9	0·9	0·9	0·9	1·6	1·6	1·6	1·6	1·6
Belgium	1·7	3·5	3·5	4·5	5·5	5·5	5·5	5·5	5·5	5·5	5·5	5·5
Brazil	—	0·6	0·6	0·6	0·6	0·6	0·6	1·9	1·9	3·1	3·1	3·1
Canada	5·2	5·2	7·0	7·8	9·3	10·1	10·8	11·6	12·5	13·4	14·2	15·1
Finland	1·3	1·3	1·3	1·3	1·3	1·3	1·3	1·3	1·3	1·3	1·3	1·3
France	12·1	19·3	22·0	25·6	30·9	35·3	44·9	47·1	47·1	46·7	46·7	45·7
Germany (F.R.)	8·6	9·8	11·0	13·4	15·8	17·0	17·0	18·5	20·0	22·6	25·0	25·0
India	0·8	1·0	1·0	1·2	1·2	1·2	1·5	1·7	1·7	1·7	1·7	1·7
Italy	1·4	1·4	1·4	1·4	1·4	1·4	1·4	1·4	2·4	3·2	3·2	3·2
Japan	14·3	14·9	16·5	16·5	18·9	22·7	26·8	26·8	27·6	29·7	29·7	29·6
Korea (S.)	0·6	0·6	0·6	1·2	1·8	2·7	3·6	4·5	5·4	5·4	5·4	5·4
Mexico	—	—	—	0·7	1·3	1·3	1·3	1·3	1·3	1·3	1·3	1·3
Netherlands	0·5	0·5	0·5	0·5	0·5	0·5	0·5	0·5	0·5	0·5	0·5	0·5
Pakistan	0·1	0·1	0·1	0·1	0·1	0·1	0·1	0·1	0·1	0·1	0·1	0·1
Philippines	—	—	—	—	—	0·6	0·6	0·6	0·6	0·6	0·6	0·6
S. Africa	—	—	0·9	1·8	1·8	1·8	1·8	1·8	1·8	1·8	1·8	1·8
Spain	1·1	2·9	3·8	6·5	7·4	7·4	8·4	9·3	11·2	12·2	12·2	11·7
Sweden	4·6	6·5	7·4	7·4	7·4	8·4	9·5	9·5	9·5	9·5	9·5	9·5
Switzerland	1·9	1·9	1·9	2·9	2·9	2·9	2·9	2·9	3·8	3·8	4·9	4·9
Taiwan	1·2	2·2	3·1	3·1	4·0	5·0	5·0	5·0	5·0	5·0	5·0	5·0
U.K.	6·6	8·5	10·3	10·3	10·3	10·3	11·6	12·4	12·4	12·0	10·3	8·6
U.S.	53·0	55·9	67·7	74·2	79·0	91·3	99·2	106·7	113·5	126·3	138·5	153·6
Yugoslavia	—	0·6	0·6	0·6	0·6	0·6	0·6	0·6	0·6	0·6	0·6	0·6
TOTAL	115·3	137·0	161·5	182·5	202·9	228·9	255·8	272·6	287·3	307·9	322·7	335·4

TABLE 2. FORECAST MOST PROBABLE NUCLEAR GENERATING CAPACITY
(GWe, net, year-end, thermal reactors)

	1980	1987	1988	1989	1990	1991	1992	1993	1994	1995
Argentina	0·3	1·6	1·6	1·6	1·6	1·6	1·6	1·6	2·2	2·2
Belgium	1·7	5·5	5·5	5·5	5·5	5·5	5·5	5·5	5·5	5·5
Brazil	—	1·9	1·9	3·1	3·1	3·1	4·3	4·3	5·5	5·5
Canada	5·2	11·6	12·5	13·4	14·2	15·1	15·1	16·3	16·3	17·6
Finland	1·3	1·3	1·3	1·3	1·3	1·3	1·3	1·3	1·3	1·3
France	12·1	48·4	51·9	55·8	59·7	62·3	64·9	67·5	70·0	72·9
Germany (F.R.)	8·6	18·5	20·0	22·6	26·5	28·3	30·0	31·2	32·4	33·5
India	0·8	1·7	1·7	2·2	2·2	2·2	2·2	2·2	2·2	2·2
Italy	1·4	1·4	2·4	3·2	3·2	4·2	5·2	5·2	5·2	6·2
Japan	14·3	31·2	34·1	36·3	38·2	41·1	45·2	49·3	52·6	54·7
Korea (S.)	0·6	4·5	5·4	5·4	6·3	7·2	8·1	9·0	10·2	11·4
Mexico	—	1·3	1·3	1·3	2·3	2·3	3·3	3·3	4·3	4·3
Netherlands	0·5	0·5	0·5	0·5	0·5	0·5	0·5	0·5	1·5	1·5
Pakistan	0·1	0·1	0·1	0·1	0·1	0·1	0·1	0·1	0·1	0·1
Philippines	—	0·6	0·6	0·6	0·6	0·6	0·6	0·6	0·6	1·2
Romania	—	0·6	0·6	0·6	0·6	1·2	1·2	1·8	1·8	1·8
S. Africa	—	1·8	1·8	1·8	1·8	1·8	1·8	1·8	3·0	3·0
Spain	1·1	9·3	11·2	12·2	12·2	12·2	12·7	12·7	13·7	14·7
Sweden	4·6	9·5	9·5	9·5	9·5	9·5	9·5	9·5	9·5	9·5
Switzerland	1·9	2·9	3·8	3·8	4·9	4·9	4·9	4·9	4·9	4·9
Taiwan	1·2	5·0	5·0	5·0	5·9	5·9	6·8	6·8	8·0	8·0
U.K.	6·6	12·4	12·4	13·1	12·5	12·3	13·4	14·1	15·2	17·4
U.S.	53·0	106·7	113·5	126·3	138·5	143·9	147·8	151·1	152·3	153·6
Yugoslavia	—	0·6	0·6	0·6	0·6	1·8	1·8	1·8	1·8	1·8
Others	—	—	—	—	0·6	0·6	1·2	1·8	2·4	5·2
TOTAL	115·3	278·9	299·2	325·8	352·4	369·5	389·0	404·2	422·5	440·0

TABLE 3. HIGH GROWTH SCENARIO NUCLEAR GENERATING CAPACITY
(GWe, net, year-end, thermal reactors)

	1980	1987	1988	1989	1990	1991	1992	1993	1994	1995
Argentina	0·3	1·6	1·6	1·6	1·6	2·2	2·2	2·2	2·8	2·8
Austria	—	0·7	0·7	0·7	0·7	0·7	0·7	0·7	0·7	0·7
Belgium	1·7	5·5	5·5	5·5	5·5	5·5	5·5	6·8	8·1	8·1
Brazil	—	1·9	1·9	3·1	3·1	3·1	4·3	4·3	5·5	5·5
Canada	5·2	12·5	13·4	14·2	15·1	15·1	16·1	17·1	18·1	19·1
Finland	1·3	1·3	1·3	1·3	1·3	1·3	1·3	2·3	2·3	2·3
France	12·1	49·7	53·6	60·1	65·3	68·0	71·0	75·1	78·9	82·6
Germany (F.R.)	8·6	18·5	20·0	22·6	26·5	28·3	30·0	31·2	32·4	33·5
India	0·8	1·7	1·7	2·2	2·2	2·7	2·7	2·7	3·2	3·2
Italy	1·4	3·4	4·4	6·2	7·2	8·2	9·2	10·2	11·2	12·2
Japan	14·3	31·2	36·3	41·8	48·1	52·1	56·2	60·3	65·8	70·8
Korea (S.)	0·6	4·5	5·4	6·3	7·2	9·0	11·4	12·6	13·8	15·0
Mexico	—	1·3	1·3	1·3	2·3	2·3	3·3	3·3	4·3	4·3
Netherlands	0·5	0·5	0·5	0·5	0·5	0·5	0·5	1·5	1·5	2·5
Pakistan	0·1	0·1	0·1	0·1	0·1	0·1	0·7	0·7	0·7	0·7
Philippines	—	0·6	0·6	0·6	0·6	1·2	1·2	1·2	1·2	1·2
Romania	—	0·6	0·6	0·6	0·6	1·2	1·2	1·8	1·8	1·8
S. Africa	—	1·8	1·8	1·8	3·0	3·0	3·0	3·0	4·2	4·2
Spain	1·1	9·3	11·2	12·2	12·2	13·2	13·7	13·7	14·7	15·7
Sweden	4·6	9·5	9·5	9·5	9·5	9·5	9·5	9·5	9·5	9·5
Switzerland	1·9	2·9	3·8	3·8	4·9	4·9	4·9	4·9	4·9	4·9
Taiwan	1·2	5·0	5·9	6·8	6·8	7·7	7·7	8·9	8·9	10·1
U.K.	6·6	12·4	12·4	13·1	12·5	12·3	13·4	15·2	17·4	20·7
U.S.	53·0	113·3	123·9	140·1	156·0	167·0	178·8	191·5	205·0	219·5
Yugoslavia	—	0·6	0·6	0·6	0·6	1·8	1·8	1·8	1·8	1·8
Others	—	—	—	—	1·3	1·3	3·5	4·1	4·7	6·8
TOTAL	115·3	290·4	318·0	356·6	394·7	422·2	453·8	486·6	523·4	559·5

the prospects for greater technology transfer, the local availability of expertise and facilities, the size of the national grid, and potential contributions from indigenous energy sources other than uranium.

In most countries it is possible to identify political actions which could by 1995 have the result of generating substantial progress towards the frequently stated aim of vigorously developing nuclear energy. However, in making the reactor capacity projections it has been assumed, for most countries, that no major political transformations will occur.

The scenario adopted for the reactivation of the U.S. nuclear programme in the *high growth* projection (Table 3) is one in which the new Administration which took office in January 1981 would make possible, by the end of its term in 1984, a shortening of lead times to periods comparable with those being achieved, for example, in France. This would permit the commissioning by late 1990 of all reactors currently on order.

After 1990, nuclear capacity in the U.S. is assumed to grow at a slightly slower rate, equal to that in the rest of the Western World under the *high growth* scenario. The *high growth* scenario for the U.S. assumes extremely rapid and far reaching changes in the operation of the U.S. Nuclear Regulatory Commission and Public Utility Commissions.

Table 4 compares the Institute's *committed, most probable* and *high growth* capacities, made on the above basis, with the latest projections of other organisations which have studied the balance of uranium supply and demand. The Institute's figures are strikingly lower than those published by the International Nuclear Fuel Cycle Evaluation, INFCE, reflecting the marked changes in perception which have taken place over the past two years.

Uranium Demand

Estimates of uranium consumption based on the Institute's three projections of installed reactor capacity — *committed, most probable* and *high growth* — are set out in Table 5 and shown graphically in Figure 2. Also shown is the feed committed under current enrichment contracts, added to the consumption of natural uranium by reactors not requiring enriched fuel. In calculating these and all other demand data presented in this report, due allowance has been made for any necessary changes in the inventory or "working stock" of uranium passing through the processing stages between mine and reactor.

Figure 2 shows that, until the mid-1980s, the feed committed to enrichment contracts exceeds the consumption of uranium by reactors using enriched fuel. In consequence, utility stocks of enriched uranium could increase over this period by around 60,000 tonnes natural uranium equivalent (tonnes U). It should be recognised, however, that a number of enrichment contracts might be terminated or adjusted, so that the reported feed requirements may overstate the uranium demand.

Beyond the mid-1980s the situation is less clear. As Figure 2 demonstrates, additional enrichment contracts will almost certainly have to be signed for that period, there being a widening gap between the enrichment feed already committed and the fuel consumption which will then be needed by operating and committed reactors. Enrichment consumers currently holding contracts for the later 1980s may, however, be given additional flexibility to vary the tails assay and hence the amount of enriched fuel produced from a given amount of feed. It is also possible that holders of enriched uranium stocks, given the higher cost of stockpiling uranium in this form, may seek to reduce inventories. It should be borne in mind that while the tables and figures illustrate the global situation, some individual consumers may have their demand for uranium determined by enrichment feed commitments throughout the period 1980–95, and others only by reactor consumption. Such differences between utilities permit them to exchange enrichment contracts and thereby to delay enrichment feed commitments where these exceed future reactor consumption.

The rate at which stocks of enriched material are consumed will clearly have a strong influence on the timing of the demand for additional enrichment capacity. A large U.S. DOE plant is currently non-operational while its

TABLE 4. COMPARISON OF NUCLEAR CAPACITY FORECASTS
(Western World, GWe net, year-end)
U.S. capacity in brackets

	URANIUM INSTITUTE (Operating and committed)	DOE/EIA (Low)	NAC (Low)	INFCE (Low)
Date of forecast	April 1981	1979 (Total) July 1980 (U.S.)	Sept. 1980	1979
1985	229 (91)	209 (86)	178 (80)	245 (100)
1990	323 (139)	292 (121)	256 (110)	373 (156)
1995	335 (154)	388		550 (200)

	URANIUM INSTITUTE (Most probable)	DOE/EIA (Mid)	NAC (Realistic)	NUKEM	NUEXCO
Date of forecast	April 1981	July 1980 (U.S.)	Sept. 1980	Apr. 1981	Nov. 1980
1985	229 (91)	(98)	205 (89)	233 (96)	207 (79)
1990	352 (139)	(128)	277 (120)	343 (130)	297 (112)
1995	440 (154)				

	URANIUM INSTITUTE (High growth)	DOE/EIA (High)	NAC (High)	INFCE (High)
Date of forecast	April 1981	1979 (Total) July 1980 (U.S.)	Sept. 1980	1979
1985	229 (91)	242 (109)	255 (112)	274 (122)
1990	395 (156)	360 (139)	376 (147)	462 (192)
1995	560 (220)	493		770 (275)

Note: Differences between the forecasts made by the various organisations can to some extent be explained by the different dates of the forecasts.

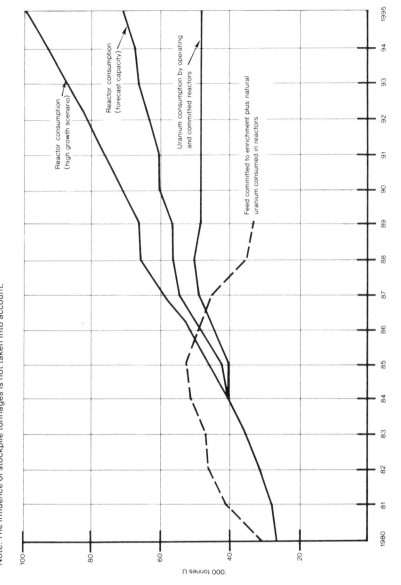

FIGURE 2. ANNUAL URANIUM DEMAND ESTIMATES, 1980–95, 000s tonnes natural U (Based on forecast installed capacity, plus *high growth* and *committed capacities* for sensitivity analysis) 0.20% tails assay 1980–85, 0.25% tails assay 1986–95.
Note: The influence of stockpile tonnages is not taken into account.

Reactor consumption (high growth scenario)

Reactor consumption (forecast capacity)

Uranium consumption by operating and committed reactors

Feed committed to enrichment plus natural uranium consumed in reactors

'000 tonnes U

13

capacity is increased; due to current over capacity, it is not clear when it will be recommissioned. New enrichment capacity may be needed after 1985. It is possible, however, that some utilities may use current overcapacity to stockpile enriched uranium, thus allowing the postponement of new construction until larger demand develops. Such a course would be technologically attractive in some cases. Thus it is not clear whether new capacity will develop in line with market demand, or whether political and strategic factors will continue to dominate the enrichment market.

Effect on Demand of Possible Variations in Operating Parameters

The demand for uranium, whether assessed on the basis of reactor consumption or by feed commitments to enrichment contracts, is subject to a considerable degree of flexibility, partly owing to the ranges within which the operating parameters of reactors and enrichment plants can be varied, and partly because the quantity of uranium purchased in any year is liable to be influenced by stockpiling policies. The possible extent of these influences is now considered.

The following are the main operating parameters and uncertainties:

Tails assay: the concentration of fissile uranium in the reject stream of an enrichment plant. This is the principal source of demand flexibility.

Load factor: the average capacity utilised in a particular year, expressed as a percentage of the design capacity. Based on past experience, load factors of 65% are assumed for light water and advanced gas-cooled reactors, while 70% is taken for natural uranium reactors (CANDU, Magnox and other heavy water or graphite-moderated designs). Load factors are affected by regulatory, as well as technical and economic considerations.

Fuel load characteristics: the quantity of uranium required to operate a given reactor, assuming a particular tails assay. This may change over time due to developments in fuel technology. For example, increasing the fuel burn-up (ie the total heat output per tonne of uranium during the residence time in the reactor) could reduce uranium consumption in the reactors concerned by up to 10%. But it should also be noted that most utilities are currently less preoccupied with saving uranium than with increasing the operating cycles of their reactors. In the case of a PWR, increasing the length of these cycles from 12 to 17 months (by means of reducing the number of refuelling periods and ensuring the availability of a greater proportion of reactors during high consumption periods) gives a net economic benefit, despite a 5–10% increase in natural uranium consumption.

Processing lead times: the delay between despatch of concentrate from the uranium mill and the loading or delivery of enriched uranium. For technical reasons, this delay can usually be defined within close limits.

Construction lead times: the effect on demand of shortening the lead times for

14

TABLE 5. URANIUM DEMAND ESTIMATES
(annual, Western World, 000s tonnes natural U, assuming no stock tonnage changes)

	1980	1981	1982	1983	1984	1985	1986	1987	1988	1989	1990	1991	1992	1993	1994	1995
A. Uranium needed to fuel *operating* and *committed* capacity (see Table 1)																
Tails assay 0·20%	27	28	31	36	40	40	41	46	47	44	44	44	44	44	44	44
0·25%	29	30	33	39	43	43	44	49	50	48	48	48	48	48	48	48
0·30%	32	33	36	42	47	47	49	54	55	53	52	52	52	52	52	52
B. Uranium needed to fuel forecast *most probable* installed capacity (see Table 2)																
Tails assay 0·20%	27	28	31	36	40	42	44	50	52	52	55	56	59	61	62	65
0·25%	29	30	33	39	43	46	48	54	56	56	60	60	63	66	67	70
0·30%	32	33	36	42	47	50	52	59	62	62	66	66	70	72	74	77
C. Uranium needed to fuel *high growth* capacity (see Table 3)																
Tails assay 0·20%	27	28	31	36	40	46	47	55	61	61	66	70	75	80	85	91
0·25%	29	30	33	39	43	50	51	59	66	66	71	76	81	87	92	99
0·30%	32	33	36	42	47	55	56	64	72	72	78	83	89	95	101	108
D. Enrichment feed for current contractual commitments, plus consumption by operating and committed natural uranium reactors																
Tails assay 0·20%	31	41	46	47	51	52	49	45	35	33						

NOTES TO ABOVE ESTIMATES:

	1980	1981	1982	1983	1984	1985	1986	1987	1988	1989	1990	1991	1992	1993	1994	1995
(i) Uranium consumption by natural uranium reactors (included above)																
Committed capacity	2	3	3	3	3	3	3	3	3	3	3	3	3	2	2	2
Forecast capacity	2	3	3	3	3	3	3	3	4	3	3	3	3	3	3	3
High growth capacity	2	3	3	3	3	3	3	4	4	4	3	3	3	3	3	4
(ii) Feed needed for current enrichment contracts (end April 1981)																
Tails assay 0·20%	29	38	43	44	48	49	46	42	32	30						

reactor construction and licensing could be dramatic: a reduction of three years in U.S. lead times before 1983 would add 10% to the Western World's uranium needs over the period 1981–90. For a given reactor capacity profile up to 1995, first core demand in the 1990–95 period would, however, be correspondingly reduced. The average lead time between commencement of construction and of commercial operation is currently ten years in the U.S. and six years elsewhere — apart from 20 reactors, all at an early stage, which may turn out to have lead times lying somewhere between the two.

The technical assumptions used in this report are set out in Appendix I, page 41.

Demand flexibilities

Demand flexibilities stem principally from the ability to vary the enrichment tails assay between 0.16% and 0.30%. (See Table 6.)

From this table it can be seen that demand can vary by as much as 28%, on changing the tails assay between 0·16% and 0·30%; a fact of considerable importance in balancing supply and demand. In the demand projections made in this report, it is assumed that use of a tails assay below 0·20% is unlikely. With an upper limit of 0·30% this still permits a 20% variation in demand.

Historically, the enrichment companies have not always allowed the utilities to choose the tails assay they may have wanted. Thus, for many years, the United States' ERDA imposed a so-called "split-tails assay" system. This distinguished between the contractual tails assay (for which the utilities had to deliver the necessary uranium feed and pay for the corresponding SWUs) and the operating tails assay (that actually used by ERDA). This policy enabled the U.S. to reduce its considerable natural uranium stockpile, substituting natural uranium for separative work. The entry of both Eurodif and Urenco into the market created competition, and ERDA (now U.S. DOE) radically revised its terms. All suppliers now give utilities the possibility of choosing, within certain limits, the tails assay they require.

Table 7 on page 17 indicates the degree of flexibility possible with each enrichment company and the periods of notice required before this can be exercised. It can be seen that, between 15 months and four years prior to delivery, utilities have the freedom to choose their tails assay. Their choice will depend very much on the anticipated prices of natural uranium and of separative work.

TABLE 6. EFFECT OF VARYING TAILS ASSAY

Change in tails assay		Effect on the natural uranium demand arising
From	to	from reactors using enriched uranium
0·20%	0·16%	− 6%
0·20%	0·25%	+ 9%
0·20%	0·30%	+20%

TABLE 7. FLEXIBILITIES IN TAILS ASSAY PERMITTED BY MAIN ENRICHMENT COMPANIES

	Eurodif	Urenco	Techsnab	U.S. DOE Requirements and Long-term Fixed Commitment	U.S. DOE Adjustable Fixed Commitment
Variable tails range	0·18–0·32%	0·20–0·30%	As set by customer, from 0·20% upwards	Not applicable	0·16%–0·30%
Notice of alteration	4 years prior to the year of delivery	4 years prior to initial delivery	9 months prior to the year of delivery	Not applicable	15 months prior to delivery
Estimated or reference tails	0·25%	As set by customer	0·20%	0·20%	0·20%

The above table shows representative tails flexibilities and periods of notice. Other terms may be negotiated, possibly subject to cost penalties. Many Techsnab customers take delivery of both tails and product, to a specification chosen by the consumer; but normal economic considerations still set a practical upper limit to the tails assay of around 0·30%.

Figure 3 relates the economically optimal tails assay to the cost of uranium, given several sets of enrichment and conversion cost assumptions. The figure shows, for example, that a utility which anticipated that it could buy uranium at \$37/lb U_3O_8, enrichment at \$125/SWU, and conversion at \$5/kg U, would optimally choose a tails assay of 0·25%, as would a utility that anticipated costs of \$27, \$100 and \$10 respectively.

The optimal tails assay, based on current prices for uranium enrichment and conversion, is about 0·27%, but the present excess of enrichment capacity has resulted in an operating tails assay close to 0·20%, thus leading to a reduction in natural uranium consumption. This excess will persist until the mid-1980s. The demand projections in Figure 2 have therefore been drawn on the basis of a 0·20% tails assay up to 1985, the period of currently perceived excess enrichment capacity. From 1986 onwards a tails assay of 0·25% is assumed. In order to allow the reader to take into account the effect of other possible tails assays, Table 5 provides annual consumption estimates on the basis of three possible tails assay levels: 0·20%, 0·25% and 0·30%.

Stockpiling Policies
Before the first energy crisis of 1973–74, few utilities thought of building up specific security-of-supply stockpiles, natural uranium being easily available in the market. The violent changes in the market during the ensuing five years, coupled with unforeseen restrictions on international trade in nuclear fuels, forced many utilities to adopt stockpiling policies.

Since 1978, a buyers' market has returned. Moreover many utilities are now carrying an involuntary stockpile over and above their normal security-of-supply targets. Few have been able to reduce their stockpiles to the intended level, either by sales of surplus fuel or by reducing contractual commitments.

Total Western utility stockpiles stood at approximately 120,000 tonnes U at the beginning of 1981. In addition to utility stockpiles, consumer Governments held a total of about 65,000 tonnes U, and intermediaries such as agents and processors held small stocks in order to maintain operational flexibility.

The target stockpiling policies adopted in various countries give an indication of the stock levels to which the utilities eventually intend to return. In practice, decisions by utilities to reduce their stocks will take into consideration a number of factors, including, in particular, the need for maintaining both short- and long-term security of supply. After 1985, most utilities are likely progressively to run down their stocks to desired levels. This policy not only involves judgements about the world political situation, but also makes it necessary to take a view of the extent to which the uranium supplying industry might be damaged.

The following country-by-country summaries, provided by Institute members, suggest that most utilities plan to hold around two years' forward

FIGURE 3. ECONOMICALLY OPTIMAL TAILS ASSAY

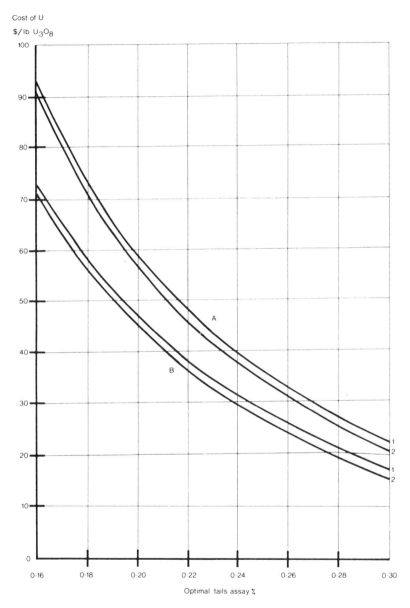

Cost of U

$/lb U₃O₈

Optimal tails assay %

Notes: Enrichment costs assumed A $ 125 / SWU
        B $ 100/SWU

Conversion costs assumed 1 $ 5/ kg U ($ 2·27/ lb U)
                          2 $ 10/kg U ($ 4·54/lb U)

requirements, although there are significant variations between national strategies. The stocks planned to be held for purposes of supply security are of course additional to the working inventories needed for the processing pipeline.

Belgian policy is that each power plant should have an inventory of two years' uranium supply.

Canadian utilities have an operating stockpile of 10 to 12 months' requirements, and the Federal Government has a range of policies designed to provide adequate security of supply. Sufficient forward mine production and reserves must be retained for each domestic reactor currently on stream, or planned to come on stream within the next ten years. The requirement is based on operation at an average annual capacity factor of 80% for 30 years (from 1980 or from the in-service date of the nuclear unit, whichever is the later). Furthermore, domestic utilities are required to maintain a contracted 15 year forward supply for both operating and committed reactors.

The policy in **France** is to keep a security stockpile sufficient to guarantee the operation of nuclear power plants for five years following a total interruption of imported supply. The stockpile target is set on the assumption that French uranium production would continue unchanged. To meet this target, EDF's stockpile should increase by the end of 1987 to the equivalent of three years' natural uranium consumption.

The **Federal Republic of Germany** currently has stocks of several thousand tonnes of uranium, mostly in natural form. This is in part the result of deliberate policy, and in part due to considerable delays in the West German nuclear programme. The majority of utilities have a stockpile target of two years' uranium needs, to be held as U_3O_8 and both natural and enriched UF_6.

There is no definitive stockpiling policy in **Italy,** but the draft nuclear programme likely to be approved by Parliament in the summer of 1981 indicates the need for arrangements to finance stockpiles. These currently stand at 7–800 tonnes of 2·75% enriched uranium, a relatively high level in terms of future demand.

There is no common uranium stockpiling policy in **Japan.** Japanese utilities procure their uranium from overseas suppliers, under long-term contracts consistent with their own installed capacity forecasts. There is no Government stockpile.

Spanish policy is to hold a minimum inventory of one year's uranium consumption, of which half will be in enriched form. In addition, Spain is building up a basic (strategic) inventory that will amount to as much as two to three years of uranium consumption. The Government will help to finance this.

Sweden intends to maintain a national stockpile of one year's forward requirements. This will be stored as UF_6.

The only stockpiling policy common to the utilities in **Switzerland** is the holding of a one year stock of fabricated fuel assemblies at the reactor site. Each utility has in addition an independently determined stockpile of natural uranium exceeding one year's consumption.

The policy of the electrical utilities in the **United Kingdom** is to maintain a stockpile of the order of two years' forward requirements, once adequate supply diversification has been achieved. In the meantime, higher stocks are considered necessary. The target stocks would be held principally in the form of uranium ore concentrates, but they would also include enriched uranium.

Inventory levels differ widely among **United States** utilities, and there is no common stock policy. However, it seems that a majority of utilities desire to maintain less than two years' forward requirements. U.S. utility stocks are very sensitive to market conditions and to the level of interest rates.

Summary on Uranium Demand

In the first half of the 1980s demand for uranium will be mainly determined by the feed necessary to service existing enrichment contracts, which, because of delays and cut-backs in the nuclear programme, are now running well ahead of real reactor needs. Demand will rise from around 40,000 tonnes U in 1981 to about 50,000 tonnes by 1985. From then on, the consumption of uranium in reactors is expected to resume its normal role as the main determinant of demand. Since the level of reactor capacity cannot be predicted with accuracy for more than a few years ahead, a range of possibilities has been considered — from a pessimistic scenario, in which no new orders are placed, through a *most probable* forecast (which is essentially based on a continuation of the present political climate) to a *high growth* forecast which, while representing a substantial shift forward from the present rate of progress, would still be well within the capability of the industry, including reactor suppliers. If the effect of possible variations in enrichment tails assay is included, uranium demand will fall within the range 45–80,000 tonnes U in 1990 and 45–110,000 tonnes in 1995. Confining attention to the *most probable* forecast, it is concluded that demand will be about 60,000 tonnes U in 1990 and 70,000 tonnes in 1995 (see Table 5, page 15).

The figures given above assume no change in the tonnage level of utilities' uranium stocks. Because of the problems facing nuclear programmes in many countries, consumers now find themselves involuntarily carrying large stocks involving substantial financial commitments. These, world-wide, amount at present to roughly $3\frac{1}{2}$ years' forward reactor consumption, appreciably more than the two years' which, on average, utilities currently regard as necessary for purposes of security of supply. The stock tonnage held does, however, represent a reasonable level for the early 1990s, when stocks will need to grow in line with reactor capacities.

21

Supply

Introduction

As developments over the past two years have demonstrated, the uncertainties involved in projecting future world uranium supply capability beyond the mid-1980s are at least as great as those inherent in forecasting demand. Any projection implies assumptions about the rate of growth of uranium requirements and the economic and socio-political climate in which the mining industry will operate; such assumptions may ultimately prove to be ill founded.

In the past, for example, the uranium industry has demonstrated an ability to respond to increases in demand by producing additional amounts of uranium at the rate required. This, however, has itself contributed to present oversupply difficulties, events of the past four years having shown that the rapid build-up of supply capacity since 1973 was based on a much higher estimated rate of growth in the nuclear power industry than has been actually realised. Whilst oversupply seems likely to characterise the uranium industry for the next few years, disruptions in major supplier countries could, however, convert a situation of oversupply to one of shortage quite rapidly. It is such unforeseeable events, together with the uncertainties over the pattern of future growth in energy requirements and the role of nuclear power, outlined earlier, which make the accurate prediction of uranium supply so difficult.

Any attempt to predict future supply capabilities requires an understanding of the level and distribution of world reserves and resources. The most widely quoted published estimates are probably those issued at two year intervals in the NEA/IAEA "Red Book". Although these provide a broad estimate of world resources, they are of no direct value to the industry for making supply projections. In practice, the basis for reserve calculations varies considerably from mine to mine, each developing a system of calculations best suited to its own particular operating or extraction method. The merging of these diverse sets of figures into national summaries and, subsequently, into a worldwide total forces difficult compromises. As the "Red Book" itself acknowledges, the reserve cost categories, as calculated by some producer countries, omit a number of incurred expenses which must necessarily be taken into account by producers when setting a selling price. The total cost of finding, mining and processing ore is the primary factor determining whether or not a mineral is an economically mineable reserve or merely of academic interest. Reserves in the ground cannot themselves satisfy demand; this can only be met by actual production from those reserves. For studies of probable future market balance, global resource evaluations are therefore intrinsically less reliable than more detailed, disaggregated studies of the *most probable* output by the producing industry.

The Institute's February 1979 report discussed, in detail, a variety of constraints and uncertainties related to longer-term uranium supply. This new report highlights a number of additional factors which are particularly relevant to the current supply situation.

Supply Capabilities

The various demand factors described earlier (pages 4–21) have contributed to reduced market prices for uranium. For reasons which will be examined later, these have resulted, in turn, in a period of instability and uncertainty which began to manifest itself in the series of mine suspensions and closures that started in 1980 and is described below. The full impact of these developments upon the production industry has yet to become clear.

Table 8 shows the pattern of uranium production in the Western World for the years 1978–80. With regard to future production, the Institute's estimates of capacity installed, under construction and in process of evaluation are shown in Tables 9, 10 and 11 respectively. These estimates were compiled from assessments made by the Supply and Demand Committee, and from data acquired by the Institute's Secretariat via contacts and correspondence with individual Institute members. To a large extent, therefore, the figures reflect aggregate data based on actual corporate plans and a detailed knowledge of each country's mining industry.

It should nonetheless be borne in mind that some of the projects listed in Table 11 are still in the early stages of evaluation. Exploration does not stop with the discovery of a deposit. Indeed, the most costly and time-consuming phase of exploration is that concerned with proving reserves after initial discovery, and gathering sufficient geotechnical data to support feasibility studies and mine design. The planning of a successful and efficient operating mine requires that prior exploration and development work is not unduly rushed, and that it proceeds carefully through all its logical stages. Much exploration work of this detailed type still remains to be done at a number of the properties included in Table 11. A degree of uncertainty must therefore attach to their grade, reserves, capital and production costs and, hence, to their overall economic feasibility.

The market conditions under which the uranium industry is operating have been marked by a number of recent changes. In real terms, spot prices peaked in 1976, representing a five-fold increase over the level prevailing three years earlier. They have declined over the past four years, however, the fall being most pronounced in 1980. Based on recently published figures, prices under long-term contracts increased in monetary terms from 1976 to 1980, but at a decreasing rate.

The period 1976–81 was one of world-wide inflation, when producers had to absorb increases in materials, mining labour, energy, and environmental control costs, which exceeded general inflation levels. For many existing producers, costs also increased as a result of a decline in the average grade of ore processed. It appears unlikely that there will be any short-term easing of these cost pressures. Inflation is forecast to remain at high levels, and many producers will experience cost increases higher than the general inflation level.

In examining future trends, it is important to note that the relationship between prices and costs is highly complex. A lengthy period of low market prices relative to production costs not only forces some higher cost mines to concentrate on higher grade areas, but also encourages all producers to become more efficient. Non-essential activities are pruned and cost saving techniques introduced. Furthermore, some pressures are placed on existing mines to extend their production capabilities in order to reduce their unit costs. Paying too much attention to today's ratio between costs and prices may carry the risk of underestimating the longer-term ability of existing producers to remain in business.

TABLE 8. URANIUM PRODUCTION (TONNES U) IN THE WESTERN WORLD 1978–80

	1978	1979	1980
U.S.	14,224	14,812	16,809
Canada	6,805	6,817	7,050
South Africa	3,961	4,800	6,100
Niger	2,063	3,629	4,505
Namibia	2,693	3,770	4,038
France	2,500	2,725	2,600
Australia	516	706	1,560
Gabon	1,022	1,100	1,000
Others	300	300	600
TOTAL	34,083	38,659	44,262

TABLE 9. PRODUCTION CAPABILITY — FACILITIES OPERATING IN 1980 (000s tonnes U per year)

	1980	1981	1982	1983	1984	1985	1986	1987	1988	1989	1990	1992	1995
Australia	1·6	2·0	2·0	2·0	2·0	1·3	1·3	1·3	1·3	1·3	1·3	—	—
Canada	7·8	8·1	8·5	8·8	8·3	8·2	8·1	8·0	7·9	7·9	7·8	7·6	3·7
France	2·6	2·2	2·2	2·2	2·2	2·2	2·2	2·2	2·2	2·2	2·2	2·2	2·2
Gabon	1·0	1·0	1·0	1·0	1·0	1·0	1·0	1·0	1·0	1·0	1·0	1·0	—
Namibia	4·0	4·0	4·0	4·0	4·0	4·0	4·0	4·0	4·0	4·0	4·0	4·0	4·0
Niger	4·4	4·4	4·4	4·4	4·4	4·4	4·4	4·4	2·2	2·2	2·2	2·2	2·2
S. Africa	6·1	6·1	6·1	6·1	5·9	5·8	5·9	5·6	5·7	5·7	5·7	5·4	5·0
Spain	0·2	0·2	0·2	0·2	0·2	0·2	0·2	0·2	0·2	0·2	0·2	—	—
U.S.	16·2	14·3	14·2	13·8	13·6	13·5	13·5	13·4	13·2	13·2	13·1	9·8	6·0
Others	0·7	0·7	0·7	0·7	0·7	0·7	0·7	0·7	0·7	0·7	0·7	0·7	0·7
TOTAL	44·4	42·8	43·1	43·0	42·1	41·1	41·1	40·6	38·2	38·2	38·0	32·7	23·4

Note: Assumes current cost/price relationships.

TABLE 10. PRODUCTION CAPABILITY — FACILITIES UNDER CONSTRUCTION IN 1980 (000s tonnes U per year)

	1981	1982	1983	1984	1985	1986	1987	1988	1989	1990	1992	1995
Australia	0·6	2·5	2·5	2·5	2·5	2·5	2·5	2·5	2·5	2·5	2·5	2·5
Brazil	—	0·5	0·5	0·5	0·5	0·5	0·5	0·5	0·5	0·5	0·5	0·5
Canada	0·7	0·9	1·0	6·4	6·3	6·3	6·3	6·3	6·3	6·3	6·4	6·4
France	0·8	1·0	1·0	1·0	1·0	1·0	1·0	1·0	1·0	1·0	1·0	0·7
Gabon	—	0·3	0·5	0·5	0·5	0·5	0·5	0·5	0·5	0·5	0·5	0·5
S. Africa	0·6	1·2	1·3	1·3	1·3	1·3	1·3	1·3	1·3	1·3	1·3	1·3
U.S.	0·8	1·0	1·0	1·0	1·0	1·0	1·0	1·0	1·0	1·0	1·0	1·0
Others	0·1	0·3	0·3	0·3	0·3	0·4	0·4	0·4	0·4	0·4	0·4	0·3
TOTAL	3·6	7·7	8·1	13·5	13·4	13·5	13·5	13·5	13·5	13·5	13·6	13·2

Note: A list of the facilities included in these tables is given in Appendix III, page 47.

However, if current market price conditions persist, the cost/price squeeze could be severe on some high-cost producers. In the medium term, and subject to pricing provisions in existing contracts, this could result in reduced output and some additional shutdowns by high cost mines, with the balance of demand being filled by lower cost mines already in operation or under construction. Positioned at the lower end of the cost range, some existing and probable producers in Canada and Australia should be well placed to take advantage of any marketing opportunities which may arise, even under current market conditions. Many of these opportunities could appear in high cost production areas, principally the United States. Much depends, however, on Government policy in the U.S., where pressure may be applied to protect the domestic uranium industry, and on the rate at which new facilities can be brought into production in Canada and Australia.

The extent to which producers are able to change their level of output in response to market conditions is considered in more detail on pages 28–30. As noted above, a number of such changes are already taking place. The period since January 1980 has seen the announcement of cuts and delays amounting respectively to some 6,000 tonnes and 5,500 tonnes of annual uranium production capacity (the word "delay" is applied to projects which were under construction at the beginning of 1980, and the tonnage given represents planned production capacity when the plant is fully commissioned).

TABLE 11. MAXIMUM PRODUCTION POTENTIAL OF PROJECTS UNDER EVALUATION IN 1980

Country	Maximum Additional Production (000s tonnes U per year)	Earliest Year of Maximum Additional Production*
Algeria	1·2	1987
Argentina	0·5	1985
Australia	18·3	1992
Brazil	0·5	1985
Canada	5·0	1991
France	1·0	1983
Italy	0·3	1986
Mexico	0·4	1983
Morocco	0·6	1992
Niger	8·1	1986
Portugal	0·2	1985
Spain	0·6	1986
Sweden	0·5	1986
U.S.	12·8	1990
Yugoslavia	0·1	1988

*This is the year in which projects in each country listed reach their aggregate maximum capacity. Some capacity will exist before this date by virtue of the build-up to full capacity at each mine, or because some mines may have reached full capacity before this time.

Note: A list of the projects included in this Table is given in Appendix III, page 47.

The country most affected by these changes is the United States, where the announced cuts are equivalent to some 30% of total production capacity at the beginning of 1980. These cuts have generally involved laying-off development rather than production personnel, so that their effect will not become apparent until 1981–82; U.S. production actually increased during 1980. Although prevailing market price conditions have clearly influenced the decisions taken by the producers concerned, it is notable that the mines affected are reported to have a wide range of production costs. This is due principally to the fact that some high cost producers are committed to existing contracts for the next few years.

It seems probable that further cuts will be announced in the near future. The way in which the pattern of supply is eventually restructured will depend on a number of considerations which include the following:

(a) the nature of producers' supply contracts, with particular regard to quantity and price. Those with firm long-term contracts containing cost escalation clauses will have a distinct advantage.

(b) the magnitude and structure of costs incurred by each producer. Those able to vary output without experiencing large fluctuations in cost, and those able to bear the cost of interim stockpiling, will have greater flexibility of choice.

(c) the producer's market objectives and strategies. Large, diversified companies may be better able to absorb high production costs during a period of low prices.

(d) political considerations, eg stockpiling policies, Government subsidies, and the imposition of import embargoes.

(e) consumer strategy, eg stockpiling and diversification policies.

(f) environmental difficulties. In some countries, notably the U.S., prohibitive costs may be incurred in meeting the stringent environmental requirements relating to the permanent closure of mines and mills, and to the reopening of those which have been temporarily closed.

(g) by-product relationships.

This restructuring will have repercussions in the medium to long-term, including the disruption of mining schedules and strategies, and a reduction in the total amount of ore which is economically recoverable. Moreover, the amount of funding available for technical research, exploration and further mine development will be limited by reduced profits and by the tendency of financial institutions to concentrate on short-term market conditions when deciding whether or not to lend for new projects. Such developments could exacerbate the uncertainties, instabilities and cyclic demand patterns experienced by all mineral industries, which were dealt with in the Institute's 1979 report.

The most important factor determining future production is *economic incentive*. In a general way, producers perceive that a long-term market, and hence some production incentive, exists for uranium; but the exact size of that market will depend on: the rate of economic growth and the associated rise in demand for electricity; on public acceptance of the need for nuclear power and the political support which must accompany it; and on the ability of electrical utilities to finance the construction of nuclear power plants. Current economic conditions in the mining industry are also important because they colour expectations for the future. In particular, current profits are partly used to finance exploration: the industry may not be able, during a period of low market prices, to take the investment decisions which, because lead times are long, are needed now in order to meet demand in the 1990s and thereafter. This could result in a shortage of expertise, resources and capital; and such changes, once established, may be difficult to reverse. The long-term stability beneficial to both producers and consumers requires that producers perceive adequate economic incentive to justify continued investment in exploration.

Short-term Flexibilities
Although the nominal capacities of uranium mines and mills are normally fixed at the design stage, there is still scope for changing the output of product by varying the mine/mill operating characteristics. This section examines the extent to which existing producers are free to increase or decrease output in response to short-term market fluctuations, and to consider the consequences of such actions in the longer term.

Increasing Production
Flexibilities for increasing production were examined in the Institute's 1979 report, and only the main points are repeated here. At the mine itself, the scope for increasing output depends largely on the nature of the mining operation. In an openpit mine there are few constraints to increasing production other than those related to the availability of men, materials, and adequate ore reserves. In an underground mine, however, severe bottlenecks may result from limited haulage, hoisting and ventilation capacity or from the use of inflexible mining methods. Shortage of skilled labour may also significantly limit the expansion of underground mines and, in all cases, remoteness of location will aggravate labour-related problems. Expansion, of course, depends on having the space available for mine waste and tailings disposal, as well as the resources to meet any associated environmental requirements.

Assuming that a sufficiently increased flow of ore is possible from the mine, three main factors will determine the ability of the associated mill to handle this material and increase its uranium output in the short-term:

Ore grade: where possible, cut-off grade may be raised in order to increase production and decrease unit costs. Excessive use of this option will, however, eventually result in severe disruption of the long-term mining schedule and a shortened mine life. Depending on the mine, this may provide a viable course of action for about three to twelve months only. In all cases, the increase in immediate output would inevitably be at the expense of some future production.

Recovery rate: mill throughput and uranium output could be increased by reducing the feed's residence time in the plant, but only at the expense of a reduced recovery rate and higher operating costs. The use of radiometric ore sorters permits increased output via an increased recovery rate. This, however, is expensive, and as often happens with solutions requiring the application of relatively new technology, is unlikely to be achieved in the short-term.

Capacity utilisation: an increase in mill output could be achieved by overtime working for short periods and by eliminating routine shutdowns and planned maintenance. However, this would again be at the expense of higher working costs, and maintenance could not be deferred indefinitely. Annual uranium output could perhaps be increased by some 5–10% for up to two years by such means. The practical significance of this source of flexibility is that it can provide a breathing space during which steps may be taken elsewhere to achieve a more easily sustained increase in output.

In general, mill capacity constraints can be eliminated more quickly than mining constraints. If labour is readily available, and if equipment manufacturers are not heavily overloaded, milling bottlenecks can — in theory at least — be eliminated within less than two years of a decision to proceed. Unfortunately, in practice, planning and environmental constraints often lengthen the effective construction period substantially.

Decreasing Production

Few technical problems are encountered in reducing mine production. Reduced production does, however, result in higher unit costs, as fixed costs are spread over a smaller output. This problem may be particularly acute in underground mines, where the cost of care and maintenance of temporarily closed sections is likely to be high.

At the mill stage, the most convenient way of decreasing output would be to reduce the feed rate. However, reducing output raises costs, and in order to keep costs down (important in a weak market and the primary reason for needing to reduce output) it would probably be necessary to raise the cut-off grade, a step which may have adverse effects on mine life and long-term mining strategy.

It is likely that the maximum reduction in output possible, while still keeping a mill operating, is about 25–30%. In some cases, however, cuts of this magnitude may not be possible, as costs may rise too high relative to price. Low-grade deposits are particularly vulnerable in this respect. Furthermore, the difficulties and adverse effects of raising cut-off grade are felt more acutely in low-grade mines, because there is less difference between cut-off grade and average grade, a fact which reduces operational flexibility. Generally the scope for reducing production is greater at high-grade mines.

Reduction of output by more than 25–30% may involve shutting down the mine and mill for a period. The cost of doing this may, however, be high, as the procedures necessary for shutdown and later reopening are expensive, and fixed costs will still be incurred during the period of closure.

The principal alternative to mill shutdown is stockpiling. Any decision for or against stockpiling will be influenced by market expectations, Government policy, the cost of production, and any by-product relationships which may exist.

It is, of course, difficult to generalise: each mine is unique and therefore subject to a different set of constraints. Nevertheless, the factors discussed above are likely to exert their influence on producers in reaching their decisions.

National Policies and their Influence upon Supply

In many countries, the regulation of mining activities by more than one level of Government has continued to increase. Nowhere is this more true than in the uranium industry where, in many areas, the net effect has been to decrease flexibility, increase lead times and raise capital and operating costs.

Disputes which have, in recent years, involved the uranium mining industry and led to Government intervention, have tended to revolve around the following issues, which are illustrated by selected examples.

Environmental impact of uranium mining and milling

Controversy over this issue is greatest in the United States, where steadily increasing pressure from various groups within and outside Government has resulted in a plethora of regulations governing the exposure of both industry workers and the general public to radiation from uranium mines and their associated facilities. The regulatory bodies have wide powers of discretion, and the cost of meeting their stringent requirements, coupled with the long delays and uncertainties associated with licensing procedures, can prove a substantial burden for existing or potential producers. The industry is conscious of the importance of maintaining high standards of pollution control. However, there must be an economic limit to the level of control which the industry can, in practice, be reasonably expected to achieve.

Land and mineral rights of local population groups

In the United States, Canada and Australia there has been conflict over land tenure and mineral rights, involving industry, local interest groups, and both local and federal authorities. The interests of local population groups vary greatly, from restriction or prevention of uranium exploration and mining to their encouragement. At one end of the spectrum, there is concern that uranium development could interfere with traditional ways of life and economic pursuits, whilst, at the other, there is the wish to ensure that local economic benefits are obtained. Such issues are, perhaps, currently of greatest importance in Australia, where the building of harmonious relationships and negotiation of land rights with Aboriginal groups set preconditions for the approval of several projects awaiting development.

Ownership and foreign participation

The two producer countries of greatest probable future importance, in which foreign participation in uranium mining ventures has been substantially restricted, are Australia and Canada. In Australia a ban on foreign participation in energy projects, imposed by the Labor Government in 1975, was subsequently replaced under the succeeding Coalition Government by a policy which limited foreign participation in uranium mining ventures to 25%. There are at present, however, no foreign investment restrictions on any uranium exploration or other nuclear fuel cycle activities. The current situation in Canada is that any individual or company may explore for uranium, but when a mine is developed to the production stage, foreign participation must be reduced to no more than 33%.

Effect of uranium imports upon the domestic mining industry

Under the low market price conditions of the 1960s, the United States Government, in order to protect the domestic U.S. mining industry, imposed an embargo on the importation of uranium for enrichment and use within the U.S. Following the improvement of the market in the mid 1970s the ban has been progressively lifted from 1977, and it is currently planned to remove all restrictions in 1984. However, with the present low market price and poor outlook for the 1980s, and with the recent closure of a number of U.S. mines, there is already some political pressure to reintroduce an embargo. The issue is not straightforward, however, and the U.S. Government would have to consider the impact which such action might have upon other aspects of international trade and goodwill.

Nuclear non-proliferation policies

The extent to which United States' producers may successfully seek foreign markets for their uranium is influenced by the insistence of the U.S. Government on full-scope safeguards for the material and by the continuing influence of the previous Administration's ban on reprocessing. In Australia, exports are also subject to full-scope safeguards under the Non Proliferation Treaty (NPT), with provision to fall back on bilateral agreements with receiving nations should the NPT cease to operate. The Australian Government recently reviewed its policy of case-by-case control over reprocessing, by defining the circumstances under which it would allow reprocessing for civil purposes to take place. The present regulations ensure that sales which have been negotiated in accordance with Government guidelines will receive approval.

Some other uranium producing nations, notably South Africa, whilst insisting on the application of IAEA safeguards or their equivalent, impose no restrictions on the internal transfer, reprocessing or recycling of the uranium which they export. Such countries may have enjoyed some market advantage in recent years, due to the fact that the case-by-case "prior consent" requirements of some other major suppliers are considered restrictive, and a source of uncertainty regarding security of supply.

The constraints outlined above are those which currently characterise the countries having the greatest potential, in terms of resources and infrastructure, to satisfy the increased demand for uranium which is likely to occur during the 1990s. Failure to ease these constraints in a period of market depression could adversely affect the prospects for a healthy, stable industry in the years to come. One consequence could be a partial diversion of funds and effort to other parts of the world where stronger investment incentives exist. While some Governments have done much to promote the future stability of the industry, to an extent which gives cause for considerable optimism, more generally Government regulation of the mining industry constitutes one of the greatest factors of uncertainty in the uranium supply/demand balance, and appears likely to remain so for some time to come.

Summary of Conclusions on Uranium Supply
Price increases in uranium during the early 1970s stimulated a considerable expansion of supply capacity and promoted an increase in the world-wide level of uranium exploration, with consequent additions to uranium reserves. The world's total supply of uranium is, however, still provided by a relatively small number of countries. They will continue to dominate the industry for the foreseeable future; but their relative rankings seem likely to change, with Canada and Australia each strengthening and improving their positions through the commitment of a number of new high-grade mines having total costs below those of a substantial proportion of existing capacity. Given its less favourable geological prospects, developments in the United States will depend largely on future Government policy and the attitude of the utilities to imports. In addition, a number of new, smaller producer countries will emerge. Although these will not make a major contribution to total world production capability, they will be of considerable strategic importance in helping to assure self-sufficiency in fuel supply at the local level. For this reason the appearance of such new capacity is likely to be relatively independent of overall market balance. Overall, the industry can confidently be expected to have the capability to meet the challenge posed by any likely nuclear programme.

32

The degree to which existing or potential reserves can contribute to future production will depend on market demand, as reflected in uranium prices, and on the impact of cost pressures on individual producers. Table 9 predicts some decline in the output of existing mines over the next decade; but their output will tend to decline much faster if profit margins shrink, because less of their potential ore reserves will then be economically exploitable. Conversely, many producing mines could sustain, or even expand, their output by developing resources previously classified as uneconomic if their profit margins were to rise. Profit margins do not depend only on the selling price but also on variations in costs, which are affected not only by external factors but also by managerial actions to improve productivity.

If the present squeeze on producers continues for an extended period, some existing capacity may be lost permanently, and some uranium ore now classified as reserves never recovered. This applies to mines which raise their cut-off grades in the face of declining profit margins, and not only to those that close. The probability of further closures is greatest in the United States, where producers' decisions are strongly influenced by the short-term costs of production. The development of operating mines can also be hindered where market pressures force reductions in labour forces, or the postponement of all but essential expenditure. This restricts their ability to respond rapidly to changing market conditions. The reopening of temporarily closed facilities can be costly, especially when subject to additional environmental constraints.

The extent to which new capacity can be created, either through the expansion of existing mines or from projects currently under construction or evaluation, depends on the demonstration of economically mineable reserves. The availability of firm sales contracts and the assurance of acceptable profitability are the essential incentives. Financial institutions, which provide most of the funds for new developments, often lay great stress on short-term market conditions when deciding whether or not to lend, although this effect can to some extent be mitigated when long-term supply contracts exist. Delays to some projects under development, and doubts about the viability of others, create risks to medium-term assurance of supply. These risks should not be unduly exaggerated, however: many potential projects seem likely to be profitable if they can secure sales contracts, while cost pressures on viable existing mines provide a strong incentive to expansion in order to lower unit costs.

In some areas, again notably the United States, the decreasing profitability of uranium mining has been accompanied by a decline in exploration activity. Finance for exploration is scarce in times of excess short-term supply and high interest rates, and competition for any limited funds which may be available is felt from other mineral commodities.

A sustained, adequate level of exploration and mine development is essential to ensure longer-term supplies. Any shortfall of production capacity from projects under construction or evaluation must be met from resources as yet unproved or uninvestigated. Bearing in mind the reservations with which one must interpret most published estimates of resources, a substantial fall in exploration efforts could have serious long-term consequences.

The balance of supply and demand

This section sums up the implications for the balance of the market of the separate supply and demand forecasts presented. At first sight, the implications for the nuclear industry are distinctly encouraging, with a three-fold growth in installed capacity likely within the next decade. The overall trend towards higher oil prices, and the progressive decrease in world oil reserves, will continue to act strongly in favour of nuclear power. However, this generally favourable outlook conceals many uncertainties now facing the industry.

These stem from factors such as the continuing upheaval in the structure of the world economy and within the energy industries themselves; high interest rates and their effect on project finance; the noticeable caution with which politicians in many countries approach questions of nuclear energy policy; and the steady proliferation of Government regulations dealing with the extraction, sale and utilisation of uranium. These influences are contributing to higher costs and longer lead times. The industry recognises that there will be no early escape from the present climate of uncertainty.

The broad shape of uranium supply and demand to the mid 1980s is relatively predictable, partly because of the long lead times needed to commission both reactors and mines. Figures 4 and 5, though excluding any contributions from stockpiles, show that annual supply is likely to exceed consumption until 1985. In response to recent market conditions a number of producers have, since the beginning of 1980, announced their intention of cutting back on existing production, to an extent which would amount to an aggregate loss of annual capacity of some 6,000 tonnes of uranium. Figure 5 shows that these cuts will help to bring annual supply and demand towards balance in the period to 1985, by decreasing the rate at which the current stockpile would otherwise have continued to grow.

It is difficult to predict on what scale further production cuts will take place. Such actions, as always, will be determined by the normal interplay of production costs, market price, supply contracts, consumer procurement policies, environmental regulations governing closures, and political considerations — as judged by individual producers operating in a free market. Without further production cuts or delays to new mines, the total utility stocks of natural and enriched uranium will represent roughly three years' forward requirements by 1985. From then onwards they could begin to decline relative to future annual consumption.

As will be seen from Figures 4 and 5, the mining industry will experience, until about 1985, a demand arising from existing enrichment contracts in excess of actual reactor requirements. In the second half of the 1980s, enrichment contracts will continue to influence primary demand, as they will determine the rate at which the utilities can run down the large stocks of enriched uranium built up during the early 1980s. In time these stocks will conform to the levels considered desirable by the utilities, whether by adjustment or as a result of increasing demand. This opens the way to a progressive rise in production capacity during the late 1980s or early 1990s. However, the extent to which demand will encourage higher production is still uncertain within a wide range. It must also be recognised that individual utility or national stockpile policies differ world-wide.

If, as anticipated, overall uranium production increases towards the end of the 1980s, this will be despite a possible fall in the production capacity of some existing mines as a result of ore-depletion or evolving reserve/price/cost ratios (see Table 9). It therefore seems probable that, soon after 1985, some of the projects shown in Table 11 as being currently under evaluation will need to be brought into production. How fast this occurs will again depend on the rate at which the nuclear industry recovers from its current recession, on the stockpiling policies of the utilities, and on the extent to which production capabilities may be found to have been adversely affected by delays and cut-backs.

The extent to which new uranium supply projects may be needed can be inferred from Table 12, which shows the annual forecast supply and demand situation for 1990 and 1995. Depending on the exact growth in demand, and on the rate of depletion of existing capacity, a substantial proportion of the total potential supply from projects currently under evaluation could be required by 1990; but the range of possibilities is wide, and at the lower limit no new projects would be needed. It is important to note that, if necessary, a significant proportion of these deposits could probably be brought into production within relatively few years, ie as fast as new nuclear stations can be planned and built. Other degrees of market flexibility include changes in enrichment tails assay, withdrawals from stockpiles, and expansion of existing mines.

36

FIGURE 4. COMPARISON OF URANIUM SUPPLY CAPABILITY AND DEMAND ESTIMATES
(Annual 1980–95, 000s tonnes U. 0·20% tails assay 1980–85 and 0·25% tails assay 1986–95)

Note: The influence of stockpile tonnages is not taken into account.

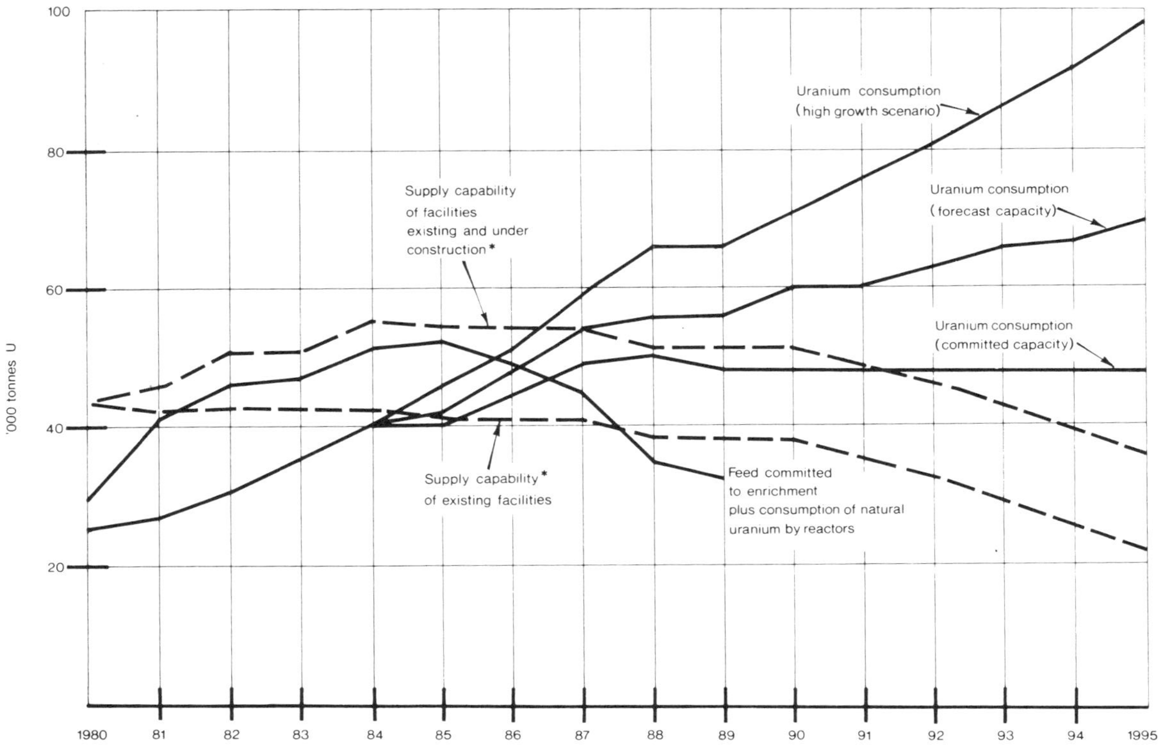

* For 1980 actual production

37

FIGURE 5. COMPARISON OF URANIUM SUPPLY AND DEMAND ESTIMATES
(Annual 1980–85, 000s tonnes U, 0·20% tails assay)

Note: The influence of stockpile tonnages is not taken into account.

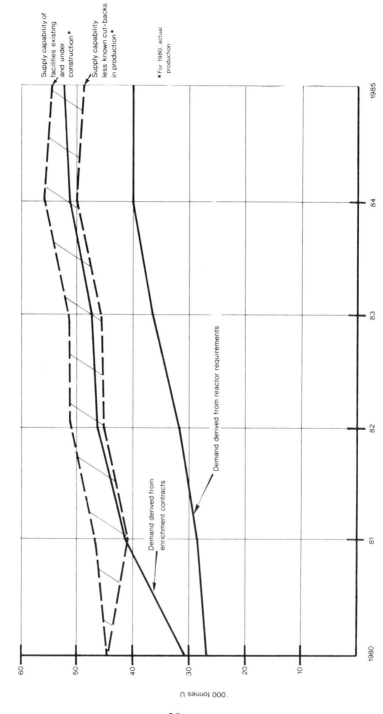

TABLE 12. THE SUPPLY AND DEMAND SITUATION IN 1990 AND IN 1995
(000s tonnes U per year, 0·25% tails assay)

	1990			1995		
	Operating and committed capacity	Forecast most probable capacity	High growth capacity	Operating and committed capacity	Forecast most probable capacity	High growth capacity
Reactor consumption (Committed/forecast/high growth capacities)	48	60	71	48	70	99
Uranium demand*	48	64	81	48	74	111
Potential supply from facilities operating and under construction		52			37	
Potential additional supply from known projects under evaluation		44			49	
Additional supply required	Nil	12	29	11	37	75
Percentage of potential additional supply from known projects required	Nil	27%	66%	22%	76%	>100%

*This Table assumes that a stockpile equivalent to two years' forward consumption will be maintained over the period considered.

On balance, therefore, there seems to be no reason to depart from the main conclusion of the February 1979 report — that the supply industry should have the capacity to meet any likely demand until after 1990.

What happens thereafter on the supply side will depend on the future course of demand and on the way in which the uranium price evolves in relation to production costs in the 1980s. If it is perceived that uranium mining offers an inadequate return on investment, particularly in comparison with other mining opportunities, this could lead to a level of expenditure on exploration and mine development insufficient to ensure adequate security of supply beyond the late 1980s. Exploration and development activity by the mining industry would, however, be encouraged by long-term purchasing commitments on the part of utilities, supplemented, in some instances, by their direct involvement as joint-venture partners.

The Institute's demand projections would require completion of some or even all the mining projects currently under evaluation (Table 11). This report has described a number of factors that could cause a proportion of these projects to fail to reach production as and when required, and reserves currently estimated for some are not yet definitive. There is, however, the possibility of production from deposits not currently under evaluation or as yet undiscovered, a possibility which is more likely to occur if a high level of exploration activity is maintained. The world stock of potential development projects is therefore something which the industry collectively needs to keep under review against the constantly changing background of future nuclear prospects.

To summarise: the present study endorses the conclusions reached in the Committee's February 1979 report, to the extent that the changes which have affected the market outlook during the past two years will not lead to a supply shortfall during the 1980s. Indeed, the present problem is one of excess supply rather than of shortage; but the forecasts suggest that, during the second half of the 1980s and the first half of the 1990s, no fundamental factors are likely to exist which would prevent supply and demand from attaining a reasonable balance. This assumes, however, that appropriate economic incentives will be available. Continued exploration and timely mine development will call for mutual understanding from both sides of the industry.

Appendix I

Technical Assumptions for Demand Forecast

Tails assay: Demand projections based on 0·20%, 0·25% and 0·30%
Load factor: 65% for PWRs, BWRs and AGRs
                70% for HWR/CANDU and GG/Magnox reactors

Processing lead times:

(a) Delay between feed   First core: 2 years
       delivery and reactor loading   Reload: 1·5 years
(b) Delay between enrichment
       delivery and reactor loading   0·75 years

Fuel requirements: (for a 1,000 MWe net reactor)

Reactor type	Requirements at 100% load factor (tonnes U, natural or enriched)	Assumed load factor	Requirements at assumed load factor and tails assay 0·20% 0·25% 0·30% (tonnes natural uranium equivalent)		
GG/Magnox					
Annual reload	306 (natural)	70%	214	214	214
AGR					
First core	165·1 (1·73% enriched)	65%	494·3	530·0	574·4
Annual reload	49·5 (2·29% enriched)		131·6	142·4	155·8
HWR/CANDU					
First core	149 (natural)	70%	149	149	149
Annual reload	170 (natural)		119	119	119
PWR					
First core	85·5 (2·36% enriched)	65%	361·4	391·3	428·5
Annual reload	35·0 (3·12% enriched)		130·0	141·6	156·1
BWR					
First core	131·4 (2·1% enriched)	65%	488·6	527·3	575·5
Annual reload	44·7 (2·6% enriched)		136·5	148·2	162·6

Reactor lifetimes: GG/Magnox 20–25 years
                        All others   30 years

Appendix II

Survey of National Reactor Programmes

As background to the data presented in Tables 1–4, brief reports follow on the status of nuclear power in countries whose consumers are represented in the Uranium Institute.

Belgium

Three commercial reactors, one of 870 MWe net, two of 400 MWe, and all of PWR type, have been operating in Belgium since 1974–75. Two 930 MWe PWRs are due to start operation by the end of 1982, and two 1,000 MWe PWRs by late 1984. For the last three years, nuclear plants have contributed 25% of electricity production.

Canada

The nuclear power programme in Canada is centred around the on-power fuelled, heavy water moderated and cooled CANDU reactor. On-power fuelling increases the availability of the units, and as a consequence annual and lifetime capacity factors have been consistently higher than those for light water reactors.

In Ontario nine reactors, with an aggregate net output of 5,250 MWe, have been integrated into the Ontario Hydro system, and an additional 12 reactors with an aggregate net output of 8,600 MWe are under construction. Two units are under construction elsewhere in Canada: the 637 MWe Gentilly 2 unit in Quebec and Point Lepreau in New Brunswick (630 MWe net).

Thus on completion of the construction programme, until recently scheduled for 1991 but now accelerated in Ontario under a directive of the Provincial Government, Canada's nuclear capacity will be 15.1 GWe net. The Premier of Ontario has also announced plans for co-generation facilities to enable the sale of steam and hot water from the Bruce nuclear complex, and in the longer term to use nuclear power in the production of hydrogen as an alternative to fossil fuels. By the end of the century Canada's nuclear capacity is expected to be in the range of 25 to 35 GWe.

42

France

At the end of 1980 there were 22 nuclear power units installed in France, with a total capacity of 14 GWe. Of these six were natural uranium/gas/graphite reactors, one a gas-cooled heavy water and one a small pressurised water reactor; 13 were 900 MWe PWRs and one the Phénix prototype fast breeder reactor.

Thirty units were then under construction: 19 900 MWe PWRs, 10 1,300 MWe PWRs and the 1,200 MWe Superphénix fast breeder reactor. These were scheduled to be commissioned by late 1986, giving a total installed capacity of 46 GWe and representing an average 5·3 GWe annual increase from 1980 to 1986.

By the end of 1981, four units should be committed (one 900 MWe PWR and three 1,300 MWe PWRs). Commissioning of these units would bring the nuclear capacity to 51 GWe net by late 1987. A capacity of 60 GWe is foreseen for the end of 1989, and 65 GWe for late 1990.

By the end of the century, an installed capacity of 85 to 105 GWe is contemplated. The commissioning of the first 1,500 MWe commercial fast breeders is scheduled to take place between 1990 and 1995. For 1988 and beyond, it is assumed that the trend established at the beginning of 1981 will be maintained although this assumption is subject to political confirmation. It has not been possible to include in this report the effect of any programme changes which may follow the election of May/June 1981.

Germany (Federal Republic)

Early in 1981, West Germany had in operation 14 nuclear power plants with a total generating capacity of 8,578 MWe net. Three of these were pilot plants of 13 to 58 MWe and 11 were commercial plants with unit sizes of between 328 and 1,240 MWe net. These commercial plants are light water reactors, six of them PWRs and five BWRs.

In 1980, nuclear plants contributed 11·7% of the total net electricity generation in West Germany and 3·6% of the total primary energy consumption in that year. This was much less than had been planned in the Federal Government's most recent Energy Programme, that of December 1977. The programme has fallen far short of official goals as a consequence of the political effects of the agitation from very active public and political minorities opposed to nuclear power.

Most of the nine nuclear power plants currently under construction (9,360 MWe net) have already suffered delays of several years, due to court actions or the late issue of partial construction permits. Four of these are 1,300 MWe PWRs and three are BWRs of the same size, the remaining two being a 300 MWe prototype THTR (thorium high temperature reactor) and a 300 MWe prototype FBR. These plants may bring the operational nuclear capacity to 17

GWe net by 1985, in contrast with the 24 GWe provided for by the Energy Programme of 1977.

Court action stopped construction of two additional 1,300 MWe PWRs, which received their first partial construction permits in 1975 and 1976. The High Court has now given the green light for one of these, the Brokdorf plant near Hamburg, and construction has begun.

The only primary energy forms suitable for additional generating capacity in the F.R.G. are coal and nuclear energy. New base load generating capacity will be needed in the future, and nuclear power has a clear cost advantage over electricity generated from coal, especially expensive domestic coal. West German utilities are obliged to maintain coal-fired capacity to burn prescribed quotas of domestic coal, but for additional generating plant they aim to add as much nuclear capacity as is politically feasible.

In addition to the two plants mentioned above, seven more 1,300 MWe PWRs are planned and licensing procedures have been started. To have all these plants in operation by 1992, bringing the total installed nuclear capacity to 30 GWe net, some improvement of the political climate would be needed; fortunately, this does not seem to be impossible.

Italy

A draft nuclear programme for Italy should receive parliamentary approval by the summer of 1981. This is contained in the National Energy Plan prepared at the end of 1980, but may be revised before presentation to Parliament. The Plan included the 1·4 GWe net nuclear generating capacity currently operating (a 1,000 MWe BWR and one small reactor each of BWR, PWR and gas/graphite designs); two 1,000 MWe BWRs now under construction and scheduled to start commercial operation by 1986–87, and a further 4 GWe of LWR capacity by 1990. Sites and reactor types have not yet been approved for these plants. An earlier nuclear programme, approved in 1977, envisaged a nuclear capacity after 1990 of up to 12 GWe and a maximum commissioning rate of 2 GWe in any year.

Japan

Japan depends for 75% of its energy requirements on imported oil. Since 1973, the considerable increase in oil prices and the experience of tight supply conditions have underlined the importance of nuclear power.

Twenty-one nuclear reactors have been installed in the 15 years since 1966, when the first came into operation, and total nuclear capacity reached 15 GWe gross at the end of 1980. Fourteen reactors are now under construction or at the planning stage, and will bring the total capacity to 32 GWe gross. The nuclear share of the total electricity generating capacity increased from 2% to 12% in the ten years to 1980.

In 1979, the Japanese Government announced its long-term forecast of energy supply. Japanese nuclear capacity was expected to reach 28–30 GWe gross by 1985, 51–53 GWe gross by 1990 and 74–78 GWe gross by 1995. The nuclear share of the total electricity generating capacity was expected to increase to 16–17% by 1985, to 22–23% by 1990 and 27–28% by 1995. The main problem in executing this schedule is that of siting, and continued efforts will be required to solve it. For this reason the figures given in Table 2, which are based on judgement, show a somewhat slower growth than the official target figures.

The majority of Japanese reactors are of light water design, and this will continue to be the case until the fast breeder reactor is put to commercial and widespread use.

South Africa
Two PWR reactors, each of 922 MWe net capacity, are under construction in South Africa. They are scheduled to start commercial operation in 1982 and 1983.

Spain
Three nuclear plants are currently operating in Spain (one PWR, one BWR and one gas/graphite), with a total capacity of 1·1 GWe net. A further seven reactors (six PWRs and one BWR), with an aggregate capacity of 6·5 GWe, are at an advanced stage of construction. One will begin commercial operation by September 1981 and the others at intervals up to 1985. The National Energy Plan for the period 1978–87 called for all these reactors to be operating by 1983.

A third group of reactors (three PWRs and two BWRs), total capacity 4·9 GWe, commenced construction between 1979 and 1981 and will come into operation in 1986–89. The National Energy Plan is now under revision, and there is no official capacity figure for 1995, but it is reasonable to assume a total of around 15 GWe.

Sweden
At present, seven reactors are in commercial operation and two in test operation, giving a total capacity of 6·4 GWe net. A national referendum in March 1980 came out in favour of a programme with 12 reactors (nine BWRs and three PWRs) and a total capacity of 9·5 GWe by 1986. This programme should be phased out by the end of the year 2010.

Switzerland
Four commercial reactors (three PWRs and one BWR) are currently in operation in Switzerland. The first three reactors installed (Beznau 1 and 2, and Muehleberg) have about the same total capacity as the first large reactor, Goesgen (920 MWe PWR), which came on line in 1979. BWR plants of

similar size are scheduled to come on line in 1983 (Leibstadt), 1988 (Kaiseraugst) and 1990 (Graben).

United Kingdom

In a statement in the House of Commons in December 1979, Mr. David Howell, Secretary of State for Energy, said that even with full exploitation of coal and conservation, and with great efforts on renewable energy sources, it would be difficult to meet the country's long-term energy needs without a size-able contribution from nuclear power.

He went on to say that the precise level of future reactor ordering would depend upon the development of electricity demand, but that for planning purposes an ordering rate of one nuclear power station a year in the decade from 1982, or a programme of 15 GWe over ten years, should be assumed. In line with this statement, contracts have been placed for two 1,300 MWe AGR stations, to be sited at Heysham, England, and Torness, Scotland, to add to the existing U.K. nuclear capacity of approximately 7 GWe in service and 4 GWe under construction.

Work towards the commercial adoption of the PWR in the U.K. is continu ing. A site for Britain's first PWR has been selected at Sizewell in Suffolk, but construction of the 1,200 MWe station cannot begin until the outcome of the public inquiry is known, and the necessary consents, nuclear licence and clearances have been given.

United States

The total nuclear generating capacity of the United States is 55,791 MWe. During 1980, 75 reactors produced 11% of the total electricity generated. Of the 69 reactors with commercial operating licences, 42 were PWRs, 24 BWRs and three were other types. Only two reactors began commercial operation during 1980, while one was retired. Three other reactors received limited licences during 1980, and it is anticipated that six will receive limited licences during 1981. A further 82 reactors have construction permits, with a combined total generation capacity of 90,503 MWe. Sixteen reactor orders were cancelled during 1980 and 15 pre-1979 orders remain. No new reactors have been ordered since 1979, the year of the Three Mile Island incident.

The new U.S. Administration generally favours nuclear power development, but a number of financial and regulatory restraints will require time to overcome.

46

Appendix III

Structure of the Uranium Mining Industry
in the Main Producer Countries*

United States

The United States is the world's largest producer of uranium, accounting in 1979 for 45% of total world output. Production for that year was distributed as follows:

Type of facility	Output tonnes U	No. of units
Conventional mills	12,945	21
In-situ leach	1,290	10
Phosphate by-product	562	4
Heap leach/tailings recovery/ copper by-product	15	4
TOTAL	14,812	

The throughput capacity of the conventional facilities totalled 41,540 tonnes ore per day. At the beginning of 1980, there were 83 producing mines owned by 31 companies. Of these companies, five controlled some 50% of total production, and 16 controlled 86% of total production. Eighty-seven companies held uranium reserves, of which 21% were accounted for by two companies, 70% by 15 companies, and 90% by 30 companies.

A striking feature of the U.S. uranium mining industry in recent years has been the steady growth of oil company involvement. Seven of the top 16 producers and six of the top ten reserve holders are oil companies. Of total reserves, oil companies control approximately 45%, including the bulk of the low-cost mineable reserves. Large, wealthy and diversified companies are better able to cope with periods of market depression, and it seems probable that the trend towards greater oil company participation in the uranium industry will continue throughout the 1980s.

---

*This appendix is included as a general reference for those not familiar with the industry.

Of the 52 known uranium deposits occurring in the United States, 43 are of the sandstone type, the remainder comprising representatives of the vein, volcanic, and igneous/metamorphic disseminated types. Of the latter, only two currently support production. These are located in Washington State and are metasedimentary deposits in which two underground mines (Midnite and Sherwood) are established, accounting for some 4% of total U.S. output. The remaining U.S. production is from sandstone deposits of Mesozoic to Tertiary age occurring, broadly, in four areas:

Area	% of total U.S. Production
New Mexico (Grants Belt)	47·0
Wyoming/E. Colorado	27·5
Texas (Gulf Coast Plain)	11·5
E. Utah/W. Colorado (Colorado Plateau)	10·0

These sandstone deposits are of three major types: "roll-front", "trend", and "stack".

Roll-front deposits predominate in Wyoming. They are sinuous or tongue-shaped in plan and crescent-shaped in vertical section, with irregular uranium mineralisation marking the contact between oxidised and unoxidised sandstone. In terms of size, the Wyoming deposits average some 750m (plan) × 50m × 5m, with grades generally in the 0·05−0·15% U_3O_8 range. Most of these deposits are shallow and are mined by openpit in production units ranging in size from 350−1,750 tonnes ore per day. Some underground and *in situ* leach mining is also carried out in Wyoming. Small roll-front deposits of simpler geometry and lower grade occur in the Gulf Coast Plain of Texas. In the past, there has been much openpit mining of these deposits, with grades averaging about 0·12−0·15% U_3O_8. More recently, however, advantage has been taken of the permeable nature of the orebodies to establish *in situ* leaching operations at locations where grades are below 0·1% U_3O_8. The bulk of production is now from facilities of this type. Small roll-front deposits, averaging some 350m × 5m × 3m and approximately 0·20−0·25% U_3O_8 in grade, also occur on the Colorado Plateau, where production is carried out by underground methods. The most important ore in this area is, however, of the trend type.

Trend ore is commonly associated with fossil plant debris, and occurs in sinuous palaeo-channels cut into the host sandstone. Orebodies are of variable size, but average perhaps 600m × 60m × 1·5m, with grades of about 0·25% U_3O_8. Production is from small underground room-and-pillar mines. Trend ore also occurs, and is exploited, in New Mexico.

Economic deposits of *stack type* ore are found in the Grants Mineral Belt of north-western New Mexico. These are so called because of the elongated vertically stacked distribution of the ore along faults and other structures unrelated to the sedimentary stratification. The ore mined ranges in grade from about 0·1−0·25% U_3O_8, and averages approximately 0·18%, having declined considerably in recent years. Most production is from shallow (to 120m), wet, underground room-and-pillar mines in the 200–1,200 tonnes ore per day capacity range, although a few mines are considerably larger than this.

Most U.S. uranium deposits are of small size, averaging some 6,000 tonnes U, and ore processing is commonly carried out on a toll basis at a conveniently located mill which handles the ore from a number of separate mines. Some U.S. producers are closely tied to domestic consumers by almost exclusive delivery arrangements.

In response to unsatisfactory rates of return resulting from market price levels and, in some cases, from increasing costs of environmental control, cutbacks and closures have been announced at 16 mining operations since the beginning of 1980. This resulted in a loss of annual production capacity of some 6,000 tonnes U, which represents approximately one-third of total U.S. capacity in 1980. In addition, nine projects currently under development were cancelled or delayed. Further cuts seem likely to follow, and the question of whether the lifting of the U.S. uranium import embargo can continue is currently under consideration by the Department of Energy.

Canada

In 1980, Canada produced a total of 7,044 tonnes of uranium, making it the world's second largest producer. Output came from seven mines. Uranium occurs in a wide range of geological environments of which only four, however, are of current commercial importance.

The bulk of present output comes from deposits of the quartz pebble conglomerate type, located in the Elliot Lake district of Ontario. These conglomerates are of Precambrian age, and are strikingly similar in appearance to the Witwatersrand "reefs" of South Africa. Four conglomerates are exploited by underground methods, at three mines, to depths of 1,000m. The ore is of great lateral extent, with thicknesses ranging from 2–4m and grades averaging some $0 \cdot 1\%$ U_3O_8. Major expansion programmes are currently under way at Elliot Lake.

The bulk of the remaining production comes from three mines located in the Precambrian Athabasca Basin of northern Saskatchewan. Deposits in this region are principally of the unconformity-vein type, many of which occur at shallow levels around the margin of the basin. The uranium occurs in irregular bodies of variable size but averaging perhaps 1,000m × 150m × 50m. Grades are generally high but variable, ranging from about $0 \cdot 4$–$4 \cdot 0\%$ U_3O_8, but with most deposits falling within the $0 \cdot 4$–$3 \cdot 0\%$ range.

Extraction is carried out in two openpit mines (Rabbit Lake and Cluff Lake), with a third openpit mine now being developed at Key Lake. Beaverlodge is an underground mine working a vein-type deposit with an average grade of $0 \cdot 2\%$ U_3O_8.

49

The remaining Canadian production comes from an underground mine in a Precambrian pegmatite at Bancroft in eastern Ontario. This has an average grade of about 0·1% U_3O_8.

Canadian production costs vary widely but much of the existing or planned production from northern Saskatchewan lies at the lower end of the cost range. The public sector has a significant involvement in Canadian uranium production at both the federal and provincial levels.

Australia

Australian uranium output in 1980 amounted to 1,500 tonnes, representing only 3·5% of total world production, making Australia the seventh largest producer in the world. However, a number of major discoveries have been made within the past 15 years, and it seems probable that Australia will emerge as a major world producer of uranium within the period under consideration in this report. Two mines are currently in operation: Mary Kathleen (Queensland) and Nabarlek (Northern Territory).

Australian uranium occurs in a variety of geological settings. The most important discoveries have been made to the east and south-east of Darwin (Northern Territory) where large deposits of the unconformity-vein type occur in rocks of Precambrian age. These are shallow occurrences, with grades in the 0·3–2·4% range, and very large reserves which, in one particular case, exceed 200,000 tonnes. A mine is presently under construction at Ranger, whilst Jabiluka and Koongarra await Government approval and feasibility studies are being undertaken on a number of other projects. The Nabarlek orebody is also of this type. Elsewhere, evaluation of sandstone-type orebodies, with grades in the 0·15–0·3% U_3O_8 range, is in progress at Westmoreland (Queensland) and at several localities in South Australia (notably Honeymoon and Beverley). Other prospects currently under evaluation include Yeelirrie (calcrete, W. Australia), Maureen (volcanic, Queensland), and Roxby Downs (disseminated igneous, S. Australia).

The deposits closest to production are those of the Northern Territory, where large, near-surface reserves and high grades will contribute to low production costs.

South Africa

With the exception of a small quantity recovered from carbonatite at the Palabora copper mine in the eastern Transvaal, most of South Africa's uranium is a by-product of the gold-mining industry. Total output (including Palabora) for 1980 was 6,000 tonnes, making South Africa the world's third largest producer.

By-product uranium is recovered from Precambrian quartz-pebble conglomerates located around the margin of the Witwatersrand basin in the southern Transvaal and Orange Free State. These conglomerates are relatively thin but of great lateral extent, with uranium values averaging some 0·015–0·02% U_3O_8. A total of 12 production centres were operating in 1980, of which nine were by-product uranium producers, and three were co-product uranium/gold producers. At one centre in each category, Chemwes and ERGO respectively, uranium is recovered exclusively from reprocessed tailings resulting from previous gold operations. A further two production centres are under construction at the present time. All mining operations are underground and reach depths locally approaching 4,000m.

Uranium ore grades vary with gold ore grades, and uranium production is therefore closely tied to that of gold. A rise in gold price could lead to a net reduction of uranium output, as lower-grade reserves tend to be mined under these market conditions.

Much recent exploration activity has focused on the search for sandstone-type deposits in the South African Karroo.

Niger
Niger is the world's fourth largest producer of uranium, with a 1980 production of 4,505 tonnes. All existing mines are located west of the Aïr Mountains in the Agadez Basin, and are established in sandstone-type orebodies of Carboniferous to Cretaceous age. Individual orebodies range in size from 25,000 to 80,000 tonnes U, with grades of about 0·10–0·45% U. There are two operating mines: one openpit (Somair) and one underground (Cominak). Three further projects are under development, and exploration is proceeding at a number of other localities. Exploration and mining is being conducted by French, German, Italian, Spanish, British, Nigerian, Japanese, Canadian and U.S. companies in association with the Niger state-owned organisation, ONAREM.

Namibia
Production in Namibia is currently confined to the Rössing mine, a large openpit operation established on a low-grade disseminated orebody in igneous and metamorphic rocks. No details of grade or reserves are published, but the grade is widely estimated to be about 0·03–0·04% U_3O_8. Production in 1980 was 4,038 tonnes, making Namibia the world's fifth largest producer.

France
France ranks as the world's sixth largest producer of uranium, with an output of 2,600 tonnes in 1980. Production is from sandstone-type and vein orebodies of variable age, which occur around the Massif Central and in the Loire valley. Deposits range in size up to 20,000 tonnes U with grades of 0·1–1·0% U. Six mines and five mills are currently operating, with two further production units under construction. Most of the orebodies are shallow, and are mined by both

openpit and underground methods. Production at individual mines is in the 100 to 1,000 tonnes U per year range.

Gabon

With a 1980 production of 1,000 tonnes, Gabon is the world's eighth largest producer of uranium. The orebodies occur within igneous and metamorphic rocks and range in grade from 0.25–0.5% U_3O_8. Current output comes from four deposits located close to Franceville, and production facilities are controlled by both Gabonese and French companies.

Appendix IV

Facilities and Projects Included in Tables 9, 10 and 11

Mines/Mills included in Table 9: Facilities operating in 1980
(Primary product, unconventional production method or other notes in brackets)

Australia
Mary Kathleen
Nabarlek

Canada
Bancroft
Beaverlodge
Calgary (phosphoric acid)
Cluff Lake
Elliot Lake (Denison)
Elliot Lake (Rio Algom)
Rabbit Lake

France
Bessines
l'Écarpière
Le Forez
Langogne
Mailhac
St-Pierre du Cantal

Gabon
Mounana

Namibia
Rössing

Niger
Akouta
Arlit

S. Africa
Blyvooruitzicht (gold)

S. Africa, *continued*
Buffelsfontein (gold)
Chemwes (gold)
ERGO (gold mine slimes gold co-product)
Harmony (gold co-product)
Hartebeestfontein (gold)
Palabora (copper)
President Brand (joint metallurgical production scheme for six gold mines)
Randfontein Estates (gold)
Vaal Reefs (gold)
West Driefontein (gold)
West Rand Consolidated (gold co-product)
Western Deep Levels (gold)

Spain
Andujar
Ciudad Rodrigo
Lobo de Laba

U.S.
Arizona
  Sahuarita (copper)

Colorado
  Canon City
  Uravan

Florida
  Boston (phosphoric acid)
  New Wales (phosphoric acid)

53

U.S. Florida, *continued*
    Pierce (phosphoric acid)
    Tampa (phosphoric acid)

Louisiana
    Uncle Sam (phosphoric acid)

New Mexico
    Ambrosia Lake
    Church Rock
    Grants
    Jackpile
    Seboyeta

Texas
    Conquista
    Panna Maria
    Pawnee (*in situ* leach)
    Zuelta (*in situ* leach)
    Palangana (*in situ* leach)
    Various other *in situ* leach

Utah
    Bingham Canyon (copper)

U.S. Utah, *continued*
    Blanding
    Lisbon
    Moab

Washington
    Midnite
    Sherwood

Wyoming
    Bear Creek
    Gas Hills (Federal American Partners)
    Gas Hills (Pathfinder)
    Gas Hills (Union Carbide)
    Irigaray (*in situ* leach)
    Jeffrey City
    Nine Mile Lake (*in situ* leach)
    Powder River
    Shirley Basin (Getty)
    Shirley Basin (Pathfinder)

Others
Unspecified

Mines/Mills included in Table 10: Facilities under construction in 1980

Australia
Ranger

Brazil
Pocos de Caldas

Canada
Elliot Lake expansion (Denison)
Elliot Lake expansion (Rio Algom)
Key Lake

France
Lodève
Mailhac expansion

Gabon
Mounana expansion

S. Africa
Beisa
Vaal Reefs expansion (gold)
Western Areas (gold)

U.S.

Utah
    Shootering Canyon

Wyoming
    Sweetwater

Others
Unspecified

Deposits in major producer countries included in Table 11: Projects under evaluation in 1980

Australia
Ben Lomond
Beverley
East Kalkaroo
Goulds Dam
Honeymoon
Jabiluka
Jabiluka expansion
Koongarra
Lake Way
Ranger expansion
Roxby Downs
Yeelirrie

Canada
Blizzard
Kitts/Michelin
McLean Lake
Midwest Lake
Milliken (Elliot Lake)

France
Bertholène
Grande Couronne
Lodève expansion

Niger
Afasto
Arni
Azelik
Imouraren

South Africa
Erfdeel-Dankbaarheid (gold)

U.S.

Arizona
　Congress Junction

Colorado
　Hansen
　Marshall Point
　Slick Rock

Florida
　Various phosphoric acid

New Mexico
　Crown Point (three projects)
　Dalton Pass
　Marquez
　Mount Taylor
　Nose Rock

S. Dakota
　Burdick

Texas
　Various *in situ* leach

Utah
　Various copper tailings heap leach

Wyoming
　Brown Ranch
　Copper Mountain
　Moore Ranch
　Morton Ranch
　North Butte Ranch
　Pumpkin Buttes
　Various *in situ* leach

55

Acknowledgement

During the preparation of this paper by the Uranium Institute's Supply and Demand Committee the following were members of the Committee. The assistance and advice afforded to the Committee by many other colleagues from member organisations is gratefully acknowledged.

Chairman: Mr. P. Erkès, Director General, Synatom, Belgium

Rapporteur: Mr. R. Janin, Assistant Director, Head of Fuels Department, Electricité de France, France

Convener of Sub-Committee on the Balance of Supply and Demand: *Mr. P. Darmayan, Pechiney Ugine Kuhlmann, France

Mr. H. Bay, Nordostschweizerische Kraftwerke AG, Switzerland
Mr. M. Beck, Comurhex, France
Mr. J. M. Bedore, Assistant Secretary-General, Uranium Institute
Mr. J. C. Blanquart, Euratom Supply Agency, Belgium
*Mr. J. Bonny, Saskatchewan Mining Development Corp., Canada
Mr. J. E. Carson, Florida Power & Light Co., U.S.
Mr. M. Chevet, Minatome S.A., France
Mr. C. Colhoun, Nukem GmbH, Federal Republic of Germany
*Mr. P. C. F. Crowson, Rio Tinto-Zinc Corp., U.K.
Dr. M. Cuzzaniti, ENEL, Italy
Dr. H. Dibbert, Rheinisch–Westfälisches Elektrizitätswerk, Federal Republic
   of Germany
Mr. I. Duncan, Western Mining Corp., Australia
*Mr. A. Genel, Electricité de France, France
Dr. P. Goldschmidt, Synatom, Belgium
Dr. H. Haas, Compagnie Francaise de Mokta, France
Mr. A. Hills, U.K. Atomic Energy Authority, U.K.
*Miss P. M. Judd, Rössing Ltd., South West Africa/Namibia
Mr. K. Kegel, Uranerz Canada Ltd., Federal Republic of Germany
*Miss A. Kidd, Rio Tinto-Zinc Corp, U.K.
*Mr. F. Klein, Urangesellschaft, Federal Republic of Germany
Mr. B. C. J. Lloyd, Pancontinental Mining Ltd., Australia
*Mr. A. J. Lorimer, British Nuclear Fuels Ltd., U.K.
*Mr. J. A. Luke, Central Electricity Generating Board, U.K.
Dr. J. Martin, Uranerz Canada Ltd., Federal Republic of Germany
Mr. H. Merlin, Uranium Canada Ltd., Canada

Acknowledgement *continued*

Dr. K. Messer, Rheinisch Westfälisches Elektrizitätswerk, Federal Republic of Germany
Mr. H. Neuman, LKAB, Sweden
Mr. B. T. Price, Secretary-General, Uranium Institute
Mr. F. Rengifo, ENUSA, Spain
Mr. J. M. van Riet Lowe, Buffelsfontein Gold Mining Co., South Africa
*Dr. E. E. N. Smith, Eldorado Nuclear Ltd., Canada
Mr. E. Strecker, Uranerz Canada Ltd., Federal Republic of Germany
Mr. M. Tsuge, Power Reactor & Nuclear Fuel Development Corp., Japan
*Mr. P. de Vaissiere, Compagnie Minière Dong-Trieu, France
Mr. R. E. Worroll, Buffelsfontein Gold Mining Co., South Africa

Secretary: Mr. W. P. Geddes, Senior Research Officer, Uranium Institute (Editor)
Assistant Secretary: Mr. J. E. Symons, Research Officer, Uranium Institute (Editor)

*Member of Sub-Committee on Balance of Supply and Demand